SLOW DEATH
BY RUBBER DUCK

SLOW DEATH
BY RUBBER DUCK

THE SECRET DANGER
OF EVERYDAY THINGS

RICK SMITH / BRUCE LOURIE WITH SARAH DOPP

COUNTERPOINT | BERKELEY

Copyright © 2009 Rick Smith and Bruce Lourie.

Published by arrangement with Knopf Canada, an imprint of the Knopf Random
Canada Publishing Group,
which is a division of Random House of Canada Limited.

Library of Congress Cataloging-in-Publication Data

Smith, Rick, 1968–
Slow death by rubber duck : the secret danger of everyday things /
Rick Smith and Bruce Lourie, with Sarah Dopp.
p. cm.
Includes bibliographical references and index.
ISBN-13: 978-1-58243-567-1
ISBN-10: 1-58243-567-7
1. Environmental toxicology—Popular works. 2. Pollution—Health
aspects. 3. Pollution—Environmental aspects. 4. Industries—
Environmental aspects. 5. Business enterprises—Environmental
aspects. I. Lourie, Bruce. II. Dopp, Sarah. III. Title.
RA1213.S65 2009b
615.9'02—dc22
2009038076

Printed in the United States of America

COUNTERPOINT
2117 Fourth Street
Suite D
Berkeley, CA 94710
www.counterpointpress.com

Distributed by Publishers Group West

10 9 8 7 6 5 4 3 2 1

For our families

CONTENTS

Slow Death by Rubber Duck takes you right to the core of one of the most alarming disasters facing us today—the invasion of man-made chemicals into every corner of our world, including our own bodies. Back in 1991, a gathering of international experts first warned us about chemicals that have the potential to disrupt the hormone systems of animals and humans. They estimated with confidence that "unless the environmental load of synthetic hormone disruptors is abated and controlled, large scale dysfunction at the population level is possible." A few years later, Pete Myers, Dianne Dumanoski, and I wrote *Our Stolen Future,* which foretold of the widespread influence of toxic chemicals on animal and human life. This early work ignited a critically important public and political debate that is still raging and to which *Slow Death by Rubber Duck* makes a substantial contribution.

When one considers that almost all of the common hormone-disrupting chemicals are derived from oil and natural gas, one can begin to understand why the public does not know the nature of these toxic chemicals, their source, and how and where they have entered our lives. The wealthiest energy corporations have long put their bottom lines before public health. So, as fossil fuel use increased, an increasing variety of chemicals were deliberately created for more and more purposes and in greater and greater volumes, and additional dollars were quietly spent to keep the public in the dark as to any problems.

As the debate surrounding climate change continues to intensify, it is important that the links between greenhouse gas emissions and other pollutants be put into proper context. Hormone disruption, like climate change, is a spin-off from society's addiction to fossil fuels. The damaging effects of hormone-disrupting chemicals on fertility, the brain and behaviour quite possibly make them a more imminent threat to humankind than climate change.

We are now into the fourth generation of people exposed to toxic chemicals from before conception through to adulthood, and statistics tell us that humankind is under siege. As a result of corporate influence over governments we now find the northern hemisphere in the midst of a pandemic of hormone-disrupting afflictions that are reaching into homes, stretching beyond the breaking point family and social service dollars, and undermining the global economy and security. A child born today faces high odds of developing at least one or more of the following ailments: attention deficit hyperactivity disorder, autism spectrum disorders, learning disabilities, diabetes, obesity, childhood and pubertal cancers, abnormal genitalia development, and infertility. Even breast and prostate cancers, and Parkinson's and Alzheimer's Diseases, have joined the above list of disorders that have been linked with prenatal exposure to toxic chemicals.

Efforts have been made to ban the sale of products for children that contain toxic chemicals. Additional steps are being taken to ban the use and continued production of specific toxic chemicals, like those the authors deliberately chose to expose themselves to and write about in this book. And government programs have been established and billions of dollars have been spent to find cures or treatments for irreversible hormone-related (chemically induced) health problems. But little or no attention has been given to the energy corporations who sell their toxic byproducts for feedstock to the companies that make the hormone-disrupting chemicals. *Slow Death by Rubber Duck* is not going to make these powerful vested interests very happy.

The most effective way to strike at the heart of the problem is to switch as soon as possible to alternative, non-fossil fuel sources of energy to reduce the availability of the basic, elemental precursors of hormone-disrupting chemicals. As an example, benzene, a toxic chemical found in coal, natural gas, and crude oil, is a key molecular building block for vast tons of hormone disruptors such as bisphenol A, phthalates, triclosan, PCBs, PBDEs, etc.—many of which are the focus of this book. The primary source of mercury (examined in Chapter 5) in the environment (and in our bodies) is the emissions from coal-burning power plants. At whatever level climate change is being dealt with—community, state or province, regional, national, or international—it should be understood that reliance on fossil fuels includes more risks than have been put on the table.

In the meantime, citizens need to be informed about the pollutants that are residing in their bodies and learn how to protect their health and the health of their families. This is where *Slow Death by Rubber Duck* can play a big role. It will take an educated citizenry to provide the necessary support and encouragement for the bold and intelligent political leadership that we so desperately need to finally put an end to pollution of all sorts.

You will find *Slow Death by Rubber Duck* difficult to put down. It is easy to read, to the point, and full of common sense. It outlines, in a very entertaining way, the challenges that we face and the steps that we need to take to protect our environment and health. *Slow Death by Rubber Duck* is surely going to become an international bestseller.

Theo Colborn
Paonia, Colorado
January 2009

In *Slow Death by Rubber Duck*, we argue that there is no separation between environmental issues and health issues. In fact, we even go so far as to say that the health of children is perhaps the most urgent environmental issue facing the United States—and the world—today. Our research conducted for this book, and the research of hundreds of medical scientists, confirms that children are most at risk to the many serious ailments linked to toxic chemicals. We refer to them as the modern childhood epidemics: asthma, autism, attention deficit hyperactivity disorder (ADHD), obesity, and reproductive disorders, among others. What is even scarier is that exposure to certain chemicals in childhood is now linked to the onset of neurological disease later in life, diseases such as Parkinson's and Alzheimer's. So the two most vulnerable populations, the young and the old, are being hurt most. And this does not even include the disease that is typically associated with toxic chemicals, cancer. According to the American Cancer Society, 1.5 million new cases of cancer are expected to be diagnosed in 2009.[1] This is truly scary stuff, and not only is it disturbing from a human health perspective, but also it places untold costs—mental and financial—on families, the medical system, and the American economy. Billions of dollars a year in medical bills can be attributed to the health effects caused by exposure to chemicals. For example, a national study estimated the environmentally attributable costs of selected illnesses and disabilities in American children at nearly $55 billion in 2002.[2] And this estimate was conservative: it did not consider the full range of illnesses and disabilities for which there is now considerable evidence of environmental causes. A California study pegged the annual costs of illness due to toxic chemical exposure to be nearly $9.5 billion.[3] New data, moreover, now implicate some environmental contaminants as causal agents in the global epidemics of type 2 diabetes and obesity. Were the portion of the costs of these diseases that may be due to contaminants included in the calculation, these estimates would be much larger.

This link between pollution and human health is a relatively new phenomenon. In fact, working on the health effects of toxic substances as recently as the early 1990s was not easy. The evidence was not well established, the medical community was sceptical at best, environmental standards focused on smoke stacks and toxic waste pipes, and chemical companies had the upper hand. Even environmental organizations suggested it was a bad idea to try to link environmental issues with health issues; "we'd lose our focus and alienate our members," said some environmentalists. But a group of smart and dedicated people were clearly on to something. The turning point for one of us (Bruce) was being invited to a dinner at a swell Washington, D.C., restaurant in February of 1995. Dr. Lynn Goldman, then EPA Special Advisor on Children's Health was the invited speaker, and she described how children are most at risk to the effects of toxic chemicals in our food, water, and air. They consume more on a body weight basis than do adults, they breathe more rapidly and therefore inhale more potentially polluted air, they crawl around poking in dusty corners and stick everything they find in their mouths. But these activities simply explain how kids have greater levels of exposure.

The most critical issues facing babies and children are that their developing bodies and brains cannot tolerate chemicals in the same way that adults can.

This dinner meeting convinced Bruce of the need to continue learning more and spreading the word. As a result he started a foundation-funding program called Environmental Contaminants and Children's Health. The program ran for nearly ten years and helped create dozens of organizations and initiatives that continue today, linking doctors, health professionals, researchers, women's health advocates, environmental groups, parenting organizations, and others, all of them working to educate people and reduce the use of toxic chemicals.

Thankfully, and in large part due to the work of the people that assembled in that Washington restaurant in 1995, it is now increasingly commonplace for Americans to be aware of chemicals they use in their homes and gardens, and there is growing public notoriety for the nasty chemicals that hide in toys, baby bottles, kids' pajamas, popcorn bags, mattresses, and thousands of other products we assume to be safe. Oprah's magazine *O* has covered toxins in everyday consumer products and even the Bisphenol A controversy in plastic baby bottles.[4] Heck, if both Oprah and Pat Robertson are talking about toxic chemicals, then we know they have arrived as an issue of major concern for Americans.[5] Manufactured synthetic chemicals are harming babies in a big way. It's as simple as that. Who in the world—other than the powerful vested interests that make money from these poorly made products—can defend this sorry state of affairs?

After World War II, the American economy boomed—oil, cars, planes, space travel, computers, and plastics. What is the common thread that ties so much of American progress together, making our world convenient yet toxic? Petrochemicals, the mainstay of plastics. The war brought incredible advances to the chemical industry, and the United States became the dominant global player in petrochemical-based plastics. "There's a great future in plastics," Dustin Hoffman's character was told in *The Graduate*—unless you are exposed to them in the womb, he failed to mention.

In 2007, the most recent year of data reported by the US EPA, 4.1 billion pounds of toxic chemicals were disposed of or released into the American environment. These are the amounts reported for almost 650 toxic chemicals monitored by the US EPA's Toxic Release Inventory (TRI).[6] The TRI database contains a wealth of information on toxic chemical releases at thousands of private and federal industrial facilities nationwide. To put this in context, this huge number refers only to the chemicals that are released into the environment in a year, ten times that quantity of chemicals, or 42 billion pounds, are produced in, or brought to, the United States *each day*.[7] These are the chemicals that wind up being used in the vast array of consumer and household products we purchase, as well as in a multitude of industrial processes.

The opening line of our book says that it is "downright hopeful," and events following the book's publication are proving this to be true. The hopefulness, however, follows a period of stagnation in the United States where the Toxic Substances Control Act, the 1976 law that gives the EPA the authority to regulate chemicals, is considered by health and environmental experts to be a legislative failure; the chemical industry on the other hand considers it to be model legislation. Kind of like a model car, one supposes. Looks nice but doesn't actually function.

All this is changing. Sophisticated organizations such as the Natural Resources Defense Council (NRDC), which monitors health and safety laws in the US, describes the "near collapse of regulatory function" for toxic chemicals.[8] The Obama administration has signalled a need to reform chemical safety laws, and with the proverbial writing on the wall, the chemical industry has come out supporting "Congress' effort to modernize our nation's chemical management system" with Ten Principles to help guide Congress.[9] Even Dow Chemical, in a backhanded way, is admitting that TSCA is not working. Dow has recently called for "enhanced regulation" with a "result that that restores public confidence" in chemicals management. Time will tell whether this seemingly new tune being whistled by the chemical industry marks a true change of heart or more of the same creative stonewalling that has characterized the industry's approach to date.

Here are a few more fascinating numbers. There are 82,000 chemicals in use in the United States with 700 new ones added each year. Of these 82,000 some odd, only 650 are monitored through TRI, only 200 have ever been tested for toxicity, and *only five* have been banned under the Toxic Substances Control Act. Not even asbestos is banned, a known carcinogen that has killed nearly 45,000 Americans over the past 30 years.[10] And finally, seven is the number dearest to our hearts, for that is the number of chemicals we write about in our book.

I suppose what captures people's attention is not only that we write about these seven chemicals, but that we expose ourselves to everyday products that contain the chemicals. And more importantly, we measure the levels as they increase in our blood and urine. The results are truly significant. But thankfully, so is the speed at which public concern is mounting and corporate and government action is starting to take place. So much of this action centers in the United States.

For keen observers of all things American, it was clear that the stories in the book would feature American companies, products, towns, scientists, and community activists. So this is a book as much about American culture as it is about anything else, and perhaps even more so about the fascinating contradictions and juxtapositions inherent in American culture. The culture of success but over-consumption, the culture of abundance but obesity, the culture fuelled by petroleum but opposed (until recently) to global warming standards, and the culture of toxic chemicals but now an emerging field of green chemistry.

Americans play the dual role of having created the toxic soup in which we bathe daily, and also of getting us out of this mess through a combination of scientific ingenuity, entrepreneurial spirit, and the diligent work of citizens, mothers, nurses, and doctors across the country. The green chemistry revolution, for example, holds huge promise. Many companies are now moving to make stuff in a non-toxic way and the Presidential Green Chemistry Challenge—an award for outstanding green chemistry technologies presented by the EPA annually—has recently recognized corporate achievements as varied as new lubricants and adhesives made out of Vitamin C instead of the current hazardous ingredients and a really promising office equipment toner made from soy instead of nasty petrochemicals.

There is also the emerging field of biomimicry where rather than creating synthetic chemicals with certain properties we desire, scientists mimic the much more elegant methods of Mother Nature to create sleek designs, water resistant properties, strong fibres, or techniques for capturing drinking water from mountain mist.

American industry is discovering that not only is there a major downside to being seen to defend toxic ingredients (witness the market share that major manufacturers of baby bottles lost as a result of being caught offside the past year's consumer backlash against bisphenol A) but that there's real money to be made in "green." Walmart's "Smart Products Initiative," in which its suppliers are being pushed to move toward non-toxic ingredients, and Clorox's new Green Works product line are just two such major corporate public relations and financial successes.

In many ways, this new and exciting public concern with toxic chemicals in consumer products is the flip side of the global warming coin. Where global warming is a huge problem necessitating ecosystem-level solutions, the clean up of toxic chemicals begins at home. The sources of our pollution are readily identifiable, and they are often innocuous household icons such as rubber ducks and baby bottles. We try in this book, with our self-experimentation, to tell the story of this very personal kind of pollution in an engaging way. We hope it builds on the rich tradition of our own pollution-fighting heroes and heroines who kicked back at corporate complacency and the nay saying status quo, allowing us to imagine a safer, toxin-free world for our children.

Bruce Lourie and Rick Smith
Toronto, Ontario
September 2009

INTRODUCTION

The four building blocks of the universe are fire, water, gravel and vinyl.

—DAVE BARRY

THE BOOK THAT YOU'RE HOLDING is downright hopeful.

Now this may seem counterintuitive, given that the word "death" appears in the title and the book describes a great many toxic chemicals that are screwing up our bodies in myriad ways. There is that. And getting all Pollyanna-ish is certainly premature.

But things can change. Sometimes very quickly and for the better.

As we wrote this book, we had to run hard just to keep up, as governments the world over complicated our writing with a European ban on noxious flame-retardant chemicals in televisions, Canadian legislative changes to put the kibosh on toxic baby bottles and, after a prolonged drought, a new U.S. law (signed by George Bush, no less) restricting hormone-mimicking ingredients in the plastic of children's toys. That's a lot of action in six months.

And as we started to catch the first glimmers of our elected leaders getting their collective act together, many people began systematically purging their homes of suspect consumer products to make way for safer alternatives.

The tide has started to turn. With surging public awareness quickly pushing the issue of toxic chemicals up the societal priority list, we set out to design something that would contribute, in some small way, to this awakening.

This is more than a book. It's kind of a big, unprecedented, adult science fair project. In the tradition of *Super Size Me* and Michael Moore, we investigated by doing. It's an unorthodox ("cuckoo," in the words of some of our loved ones) and very personal examination of the chemicals in our own bodies and the lives of our families. Along the way we've confronted the companies that made the chemicals, interviewed the government regulators who looked the other way while problems mounted and met the scientists and community organizers who are making a difference.

In our day jobs we're long-standing environmental advocates in Canada. We toil away in the trenches, trying to secure better government policy to protect the environment and human health. The idea for this book came out of that work, and specifically from Environmental Defence Canada's Toxic Nation project, a campaign to expose the dangers of pollution through testing Canadians for measurable levels of toxic chemicals in their bodies.

A New Kind of Pollution

Far from being the rock or island in the Simon and Garfunkel song, it turns out that the best metaphor to describe the human body is "sponge." We're permeable. We're absorbent. And Toxic Nation tries to measure the nasty things the human sponge has soaked up. Like efforts in the United States and Europe, the Toxic Nation project applies scientific testing techniques—previously restricted to the pages of obscure scientific journals—to the raging public debate about what pollutants we are exposed to, in what amounts and from which sources—and tells us what we can do about it. Since 2005 Environmental Defence Canada has tested the blood and urine of more than 40 Canadians for over 130 pollutants. People from all walks of life. Of all ages. Men, women and kids from different parts of the country and different ethnic backgrounds. They all turned out to be polluted to some degree.

As we chatted about the implications of these findings with the test volunteers, the media covering the story and the members of the public who took notice, it became clear that the whole concept of "pollution" that we carry around in our heads needed updating.

Belching smokestacks. Sewer outfalls. Car exhaust. For most people these are the first images that come to mind when the word "pollution" is mentioned. It's still seen as an external concern. Something floating around in the air or in the nearest lake. Out there. Something that can still be avoided.

As our Toxic Nation testing makes clear, however, the reality is quite different. Pollution is now so pervasive that it's become a marinade in which we all bathe every day. Pollution is actually inside us all. It's seeped into our bodies. And in many cases, once in, it's impossible to get out.

Baby bottles. Deodorants. A favourite overstuffed sofa. These items, so familiar and apparently harmless, are now sources of pollution at least as serious as the more industrial-grade varieties described above. The market-leading baby bottles in North America are made of polycarbonate plastic, and they leach bisphenol A, a known hormone disruptor, into their contents. Deodorants—and nearly every other common product in the bathroom—can contain phthalates (pronounced "tha-lates"), which have been linked to a number of serious reproductive problems. Phthalates are also a common ingredient of vinyl children's toys. Sofas and other upholstered products contain brominated flame retardants and are coated with stain-repellent chemicals, both of which increase the risk of cancer and are absorbed by anyone sitting on a sofa or chair to watch Friday night TV.

We found all of these chemicals, and many more, in the bodies of the Canadians we tested.

The truth of the matter is that toxic chemicals are now found at low levels in countless applications, in everything from personal-care products and cooking pots and pans to electronics, furniture, clothing, building materials and children's toys. They make their way into our bodies through our food, air and water. From the moment we get up from a good night's sleep under wrinkle-resistant sheets (which are treated with the known carcinogen formaldehyde) to the time we go to bed at night after a snack of microwave popcorn (the interior of the bag being coated with an indestructible chemical that builds up in our bodies), pollution surrounds us.

Far from escaping it when we shut our front door at night, we've unwittingly welcomed these toxins into our homes in countless ways. In a particularly graphic example, it's been estimated that by the time the average woman grabs her morning coffee, she has applied 126 different chemicals in 12 different products to her face, body and hair.

And the result? Not surprisingly, a large and growing body of scientific research links exposure to toxic chemicals to many ailments that plague people, including several forms of cancer, reproductive problems and birth defects, respiratory illnesses such as asthma and neurodevelopmental disorders such as attention deficit hyperactivity disorder (ADHD).

We have all become guinea pigs in a vast and uncontrolled experiment.

At this moment in history, the image conjured up by the word "pollution" is just as properly an innocent rubber duck as it is a giant smokestack. The first chapter of this book makes this case by giving a whirlwind history of pollution and examining how humanity's ability to poison itself has changed from a local, highly visible and acute phenomenon to a global, largely invisible and chronic threat. A threat very often coming from everyday household products.

Cause and Effect

Another insight that came to us through Toxic Nation is that once people realize they're immersed in pollution, it's a fine line between motivating them to action and having them lapse into a kind of pollution nihilism. "If it's all around us, there's not much I can do, is there?" is a comment we heard frequently throughout the Toxic Nation project.

The need for specific answers was something that very much preoccupied the Toxic Nation test volunteers, regardless of whether it was Canada's Minister of Health (one of a few politicians who let us draw their blood) or a ten-year-old kid from Montreal. The first question they all asked upon seeing their results was "How did this pollution get into me?" Talking in generalities about pathways of exposure (e.g., "This chemical is commonly found in plastics; this one is generally in upholstered products") wasn't enough to satisfy their curiosity. They wanted to know what act, on what day, had led to this level of pollutant in their blood. They wanted some assurance that if they started making different choices, such as buying more environmentally friendly personal-care products, they would see a decrease in their pollution levels. In short, they wanted an explanation of cause and effect that, in many cases, we were unable to provide because the studies hadn't yet been done.

For example, we could tell them that researchers in Denmark have demonstrated that rubbing a laboratory preparation of phthalates over the entire body resulted in increased phthalates in the urine. But this doesn't help much in the real world. Phthalate levels aren't marked on shampoos or other off-the-shelf products. If you're lucky the word "Fragrance" on the fine-print ingredient list is an occasional tip-off as to their presence. Would the normal use of name-brand personal-care products really affect someone's phthalate levels?

"Probably" was the best answer we could muster.

For some chemicals, like bisphenol A (BPA), there are virtually no human data available at all. Nobody had ever tried to raise and lower BPA levels in a person's body before. So telling people to stop microwaving their leftovers in polycarbonate containers because it would expose them to chemicals that leach out of the plastic felt just the tiniest bit wobbly in terms of certain outcomes.

As we talked about how to answer the questions outlined above, the germ of an idea started to take shape.

Only One Rule

"Why don't we experiment on ourselves?"

What began as a joke, an offhand thought, quickly became a two-year megaproject. The more we chewed it over, the more doable it seemed. What better way to demonstrate, in concrete terms, the impact of daily life on the pollution load our bodies all carry than to deliberately ingest a whole bunch of these suspect substances and see whether they did, in fact, linger in our systems?

We set only one ironclad rule: Our efforts had to mimic real life. This may seem obvious, but it was actually a very useful guiding principle as we wrestled with the details of the experimentation. We couldn't chug a bottle of mercury. We couldn't douse ourselves in Teflon. Whatever activities we undertook had to be run-of-the-mill things that people do every day.

As we started consulting experts and poring over scientific studies, it frequently felt as if we were assembling a giant puzzle. The critical pieces that needed fitting together were a list of chemicals for which there was mounting human health concern, a good sense of daily activities that might expose the average person to these chemicals and the outline of an experiment that would reveal whether these daily activities measurably affect the levels of the chemical in question in our bodies.

We measured any increase or decrease in ourselves by methodically taking blood and urine samples before and after performing the activities. After considering many different options, we decided to take a look at seven toxic chemicals and divided up the chapters so we could tell the stories in the first person. This self-experimentation with dicey toxic chemicals, which so delighted our families (not!), was an experience best shared, we figured.

In Chapter 2 Rick experiments with phthalates and sets out to get some answers from the toy industry, which seems intent on poisoning his kids. Bruce picks up the story in Chapter 3 by taking a trip to Parkersburg, West Virginia, the town that Teflon built, to see what happens when a company invents a chemical that lasts forever. In Chapter 4 Rick travels to Victoria, British Columbia, to speak with experts about the "*déjà vu* all over again" of brominated flame retardants, a family of compounds that seems to be repeating the nasty history of PCBs. Bruce, in Chapter 5, then gives a very personal account of mercury, the oldest known toxin. In Chapter 6 Rick successfully cranks his levels of the antibacterial chemical triclosan into the stratosphere and asks why we're so terribly afraid of germs. In Chapter 7 Bruce confronts head-on the way in which the chemical industry continually asks us to assume risk so they can make more money. And in Chapter 8 Rick cooks with plastic and outlines how moms and dads are confronting the chemical industry when it comes to bisphenol A. Our über-organized project coordinator, Sarah, was the glue that held the whole effort together. She dealt with the complicated logistics of the experiments, blood and urine testing and communications with the laboratories. She assembled the masses of sometimes difficult-to-find research upon which the book is based.

The book concludes with a road map showing how simple changes in consumer choices can detox our lives and how the average citizen can help twist the arms of elected leaders so they'll do better in protecting us from these toxins.

We won't be surprised if this book annoys the pro–chemical industry, anti-environmental pundits who think or pretend (we're not sure which is worse) that nothing in society should be regulated without absolute scientific certainty. These writers and lobbyists like to call our work and the work of any other scientists who identify health problems linked with synthetic chemicals "junk science." The tests we carried out for the book follow standard science protocols and they're easily replicable. Though they do not include large sample sizes, double-blind trials or other methods that constitute formal scientific research, what matters is that they demonstrate the surprising reality that a couple of guys can manipulate the toxic substances in their bodies through the simple acts of eating and using everyday foods and products.

For readers wishing to understand some of the crazy ideas behind the proliferation of toxins around the world, we hope that this book will shed some new light on the issues. A light that is too often obscured by chemical companies and their batallions of hired-gun consultants, industry-funded academics and con-flicted government bureaucrats.

As Rachel Carson wrote in *Silent Spring:* "For the first time in the history of the world, every human being is now subjected to contact with dangerous chemicals, from the moment of concep-tion until death." That was 1962. Let's see how we're doing today.

ONE: POLLUTION THEN AND NOW

Put the argument into a concrete shape, into an image, some hard phrase, round and solid as a ball, which they can see and handle and carry home with them, and the cause is half won.

—RALPH WALDO EMERSON, *Society and Solitude*, 1870

KEN COOK REMEMBERS the exact moment he had the idea.

"It was 1998 and I was riding my bike around Hains Point in Washington, D.C., a little spit of land between the Potomac and Anacostia Rivers. It's very flat, and you can go all out. It's a good time to think because there's no traffic and no distractions. And that's when it popped into my mind."

Energetic, with an infectious—almost boyish—enthusiasm, it's somehow not surprising that Cook does a lot of thinking on his bicycle. For years Cook and his colleagues at the Environmental Working Group (EWG) had been at the forefront of the pollution debate in the United States, pioneering direct measurement of pollution levels as a way of tangibly highlighting the problem. "We had been testing for pollution in air and water and food and consumer products for some time. We were very good at it. But there was always the question from the chemical industry of 'Well, yeah, it may be in the air, it may be in the water but, honestly, are people really being exposed in a significant way?'" Though this was obviously "just a dodge" on the part of industry, Cook

9

and his colleagues were tired of not having the data to rebut this maddening "don't worry, be happy" argument.

What to do?

As he pedalled his bike and watched the water lazily flowing by and the planes flying into National Airport, Cook had an epiphany that would redefine the pollution debate in the United States and around the world:

What if people started finding out what was in them?

What if EWG focused its testing on the bodies of Americans as opposed to the external environment?

Recounting the moment it all came together for him, Cook says he thought of a 1970s newspaper ad by the U.S. Environmental Defense Fund (EDF) that linked pesticides with human breastmilk. "And then I remembered this passage from *Our Stolen Future* saying that virtually anyone willing to put up a couple of thousand bucks for the tests will find at least 250 chemical contaminants in their body fat."

Released in early 1996, the best-selling book *Our Stolen Future*[1] has been dubbed (by Al Gore) the sequel to Rachel Carson's *Silent Spring*. While Carson's book dealt mostly with the effects of pesticides on birds and wildlife, *Our Stolen Future* rang the alarm bell on a new pollution concern: hormone disruption, the damage done to human sexual development and reproduction by the ocean of synthetic chemicals produced by industry each year. The ability to detect these chemicals was a relatively recent phenomenon—the product of rapid technological advances in laboratory testing methods—and the prevalence of this new kind of pollution took many people by surprise.

When he returned to the office that Monday, Cook and his staff started to put a plan together. They decided to start out small. Cook himself and *Our Stolen Future* co-author Pete Myers "were the first guinea pigs" to donate their blood and urine for laboratory analyses. The trial run worked. Though it "turned out the cost estimate of the testing was a little off, the basic fact was there; the measurement of pollution in people was a possibility," Cook recalls.

As EWG kept working out the kinks in their test protocols, the next batch of volunteers were personal friends and acquaintances of EWG staff. It was still too novel and disturbing a request to make of strangers: "Can you please give us some blood and pee so we can tell you how many toxic chemicals you have in your body?" Though the notion of testing people for chemical contamination was not new (for instance, as early as the 1890s, factory workers who were exposed to lead had their blood and urine screened to enable early detection of lead poisoning), EWG's approach was innovative for a couple of reasons.

First, the organization was testing for a very large number of chemicals at once: their first report found 171 of 214 possible toxic chemicals in the volunteers' bodies. They tested for pesticides, heavy metals, perfluorinated compounds (PFCs) (chemicals similar to Teflon), brominated flame retardants, you name it. The aim was to describe the smorgasbord of pollutants in our veins and to make these results public to drive the debate forward. This was science harnessed to policy advocacy in a brand-new way. It necessitated the creation by EWG of a whole new lexicon to describe the toxic chemical load we all carry: "body burden," "biomonitoring" and the most evocative term, "human toxome"—created in deliberate contrast with the "human genome" that scientists are so intensely interested in mapping.

The other EWG innovation was to cast aside the anonymity of the study participants—usually a staple of human experimentation. In this case the whole point was to be public. To recruit volunteers who were willing to discuss their "body burdens": a pollution confessional for the reality TV age. Bill Moyers, the famous PBS journalist, was one of the first people tested by EWG and the first to publicly discuss his results on air in 2001.

As we will see later in this chapter, he was only the first of many who stepped forward to literally become the new face of pollution.

It's not often that a single decision totally changes the dynamic of a major global public debate. Ken Cook's bike-borne insight did exactly that. The detection of hundreds of toxic chemicals in people made it crystal clear that in the words of *Our Stolen Future,* "regardless of whether [you live] in Gary, Indiana, or on a remote island in the South Pacific . . . [you] cannot escape them."

Updating Dr. Seuss

The chemical industry didn't see the body burden testing coming, and they don't like it one little bit.

"It's very hard for them to defend themselves. How do you say, 'Just a little bit of my company's chemical is in your baby's blood while your baby is still in the womb, and even though it's not proven safe, it's really nothing to worry about, madam'?" says Ken Cook in summarizing the chemical industry's challenge. But even though their arguments are now harder to spin, the industry and their friends are still trying hard.

To listen to them you'd think that pollution is a problem on the verge of being licked.

Bjørn Lomborg, author of the controversial 2001 book *The Skeptical Environmentalist,* presents a case that air pollution is not a new phenomenon that is getting worse but an old phenomenon that has been getting better, leaving London, England, cleaner now than it has been since the Middle Ages. The American Chemistry Council, the chemical industry's primary lobby group in the United States, relentlessly sings the praises of their voluntary Responsible Care program and its supposed effectiveness in reducing chemical releases from its member companies. And Elizabeth Whelan, president of the American Council on Science and Health (which generally takes an apologetic stance where industry is concerned), thinks that carcinogens in the environment are not the problem they've been made out to be. This is just a small sampling of industry's general mantra that pollution doesn't hurt us that much, isn't as bad as we think and is much better than it used to be.

To give the chemical industry its due, it is true that some pollution, in some places, has been cleaned up.

The Thames River running through London, England, is famous among these successes. By the end of the 13th century, it was already known to be polluted. Some six hundred years later, in 1834, in what is thought to be the first successful indictment of a company brought forward on nuisance charges, the City of London filed suit against a coal gas manufacturer for releasing coal tar into the river.[2] Throughout the 19th century the Thames was further transformed into the city's collective cesspool, and by the 1950s the water quality was so poor that the river was declared biologically dead. Amazingly, through a concerted government and public effort, by 2000 there were renewed signs of life, and in 2007 the Thames was considered clean enough to reintroduce salmon for the first time in centuries.[3]

The death and resurrection of Lake Erie is another example of things changing for the better. Industrial waste became an issue in the Great Lakes in the 1950s, and by the 1960s Lake Erie was declared dead. "Eutrophication had claimed Lake Erie and excessive algae became the dominant plant species, covering beaches in slimy moss and killing off native aquatic species by soaking up all of the oxygen."[4] Yet after much work on both sides of the border, by the late 1970s the International Joint Commission (IJC) issued a report providing evidence that chemical pollution had waned and local gull populations were recovering.[5]

Lake Erie's resurrection was astonishing enough that it resulted in edits to one of the timeless classics of children's literature. *The Lorax* by Dr. Seuss was based on the state of pollution in the United States in 1971. It's a tale of environmental warning and features the Once-ler—a greedy character who cuts down Truffula trees so he can use the silk tufts to knit highly lucrative Thneeds. Thneed sales are so successful that he builds a factory and invents a machine to cut down four trees at a time. The Lorax, a mossy, tree-dwelling creature that looks like a cross between Santa Claus and Oscar the Grouch, speaks out to defend the trees and the ecosystem of which they are part. But the Once-ler will not be deterred. He continues until the last tree is cut, and his production comes to an abrupt end. The environment in which the Once-ler and the Lorax live is left barren and polluted.

In the original text Seuss included the line "I hear things are just as bad up in Lake Erie." Fourteen years after the book was published, Seuss was contacted by two scientists with the Ohio Sea Grant program, who updated him on the success of the Lake Erie clean-up. While the line was removed from subsequent copies of the text, it remains in the DVD release of the TV special.[6]

With respect to air pollution, a variety of indicators point to improvements over the past few decades. Since the U.S. passed its Clean Air Act in 1970, major pollutants have decreased 48 per cent (though the health effects of air pollution continue to afflict thousands of Americans).[7] Similarly, after the U.K. passed its Clean Air Act in 1956, introducing controls on the types of fuel that could be burned in the city centre, the amount of smoke in the air over London fell by 80 per cent over the subsequent 15 years.[8]

Encouraging? Yes. Evidence that humanity's pollution difficulties are at an end? No. Pollution simply looks different than it used to. It's changed form. Many of the chemicals EWG tests for, and all of the toxic chemicals we experiment with in this book, are now much more common and present a much greater threat to human health than at any time in the past.

Though admittedly a generalization, it seems to us that over the past few decades, pollution has changed dramatically in the following important ways:

1. It's now *global rather than local*
2. It's moved from being *highly visible to being invisible*
3. In many cases its effects are now *chronic and long-term rather than acute and immediate*

Let's look at a few examples of how these shifts have occurred and what this has meant for the toxics debate.

The Broad Street Pump

Even the caves of our earliest ancestors experienced pollution problems. It's always been a challenge for humanity to know how to dispose of waste, whether the bodily variety or the other unwanted byproducts of living.

Agriculture in ancient Sumeria in the third millennium B.C.E. was plagued with salt buildup. By 2100 B.C.E. salt pollution from bad agricultural practices and soil erosion had devastated the fields, prompting one Sumerian to write that the "earth turned white."[9] The ancient Greeks created what is considered to be the first municipal dump in 500 B.C.E., requiring garbage to be disposed of at least a mile from the city walls.[10]

The links between pollution and human health have been known for centuries. As far back as the 10th to 12th centuries C.E., pollution and health were written of extensively in Arabic medical treatises.[11] Many of these papers were concerned with air and water pollution in concentrated areas like Baghdad, Damascus and Cairo. One paper describes the use of incense to purge the air of spoilage, while another examines the "positioning of dwelling units uphill and upwind from infected areas."[12] The writings of Ibn Sīnā describe the treatment of illnesses caused by impure water and the creation of pollution by animals and their waste.

For most of our history, pollution was a highly localized, very visible (or smelly) and often deadly phenomenon. No example better illustrates this than the story of the Broad Street pump. In the Soho district of London in 1854, a cholera outbreak ripped through the community in the most sudden of ways. It wasn't the first epidemic, nor would it be the last, but in just ten days over five hundred incidents of cholera were experienced within a few city blocks. It was the investigative work of Dr. John Snow that uncovered the mysterious source of the outbreak. Snow had been developing his theory of water-borne contamination as a cause of cholera since an earlier and deadlier outbreak of the disease in England in 1848–49 that had claimed fifty thousand lives.[13] After looking at the pattern of the Broad Street infections, Snow pinpointed the epicentre as the neighbourhood pump. He was able to compel local authorities with his evidence, and the handle of the pump was removed, thereby effectively stopping the spread. It was later determined that the pump's contamination was due to the recent disposal of human waste in the area.

The Broad Street pump incident was highly local: the epidemic touched very few beyond Soho's boundaries. In fact, those outside the neighbourhood who were affected had direct connections by way of family who lived or worked in Soho and had their drinking water brought to them from the Broad Street pump. While the stench intensified as the cholera spread, the physical, visible effects of the disease manifested themselves just as quickly. Sunken eyes, a bluish tinge to the lips and dramatic weight loss within a matter of hours were noted among those who were felled. Bodies were carried by the cartload through the streets.[14]

Snow did not live to see the cause of cholera publicly accepted; however, his work—and the Broad Street incident—contributed to the establishment of sewer systems in London, a model followed by other Western cities as they cleaned up their waterways and improved public health.

Dead Rivers, Killer Fogs

For most of human history, pollution was very much in your face.

Chronicles from the time of the Industrial Revolution, for example, pull no punches. The cities of England at this time were a putrid, stinking mess. In *Bleak House* Charles Dickens talks about the November "smoke lowering down from the chimney-pots, making a soft black drizzle, with flakes of soot in it as big as full-grown snowflakes—gone into mourning, one might imagine, for the death of the sun." Friedrich Engels described the River Irk in Manchester in the 1840s as "a coal-black stinking river full of filth and garbage which it deposits on the lower-lying right bank. In dry weather, an extended series of the most revolting blackish green pools of slime

remain standing on this bank, out of whose depths bubbles of miasmatic gases constantly rise and give forth a stench that is unbearable
even on the bridge forty or fifty feet above the level of the water." The
term "acid rain" was actually coined in 1852 by the Scottish chemist
Robert Angus Smith to describe the link between the Manchester
region's polluted skies and the acidity of its rainfall.

Well into the 20th century, lakes and rivers remained terribly
polluted throughout the Western world. One of the most spectacular symptoms of their plight was the fact that they would periodi-
cally catch on fire. One such incident happened in June 1969 on
the Cuyahoga River in Cleveland, Ohio. Oil and chemical pollution
fuelled the flames that reached a height of five storeys. And this
wasn't the only time the Cuyahoga had been ablaze. The largest fire
on the river had occurred in 1952, causing over one million dollars
in damage. In just five years, between 1965 and 1970, chemical
pollution caused fires on the Iset River in Sverdlovsk (known today
as Ekaterinburg) and the Volga River in the former Soviet Union.
In both cases pesticides were a significant component of the combustible mixture. In yet another unusual incident, a fire at a chemical plant in Switzerland released 30 tonnes of pesticides into the
Rhine River, turning the waterway red.[15]

Air pollution in the first half of the 20th century was sometimes
shockingly bad. For example, in December 1930 the industrial
valley of the Meuse River in Belgium was plagued by a heavy, chok-
ing fog. This incident marked the first time that a link between air
pollution and disease was scientifically determined. Over three
days 60 people lost their lives, their deaths directly attributable to
the fog. A committee formed to investigate the events came to the
conclusion that sulphur, produced by burning coal, was the culprit
and that sulphur–compound pollution had increased by as much as
tenfold as a result of increased population and industrialization.[16]

It happened in America too. In 1949 a thick fog of pollution
settled over Donora, a mining town in Pennsylvania. For four days
the fog hung over the town, bringing near darkness and, ultimately,

death. Twenty people perished during those three days, and about six thousand became seriously ill over the ensuing months.[17]

Five years later London, England, was also engulfed in what has become known as the Great Smog of 1952. ("Smog" was a term coined early in the century as a contraction of "smoke" and "fog.") More than four thousand people died in a few short days, and a further eight thousand perished from its effects during the following weeks and months. The incident occurred after a particularly cold period, when people had been burning more coal than usual. Thousands of tonnes of black soot, tar particles and sulphur dioxide accumulated in the air, creating a dense fog in the city streets.[18] This incident led directly to the establishment of the U.K. Clean Air Act in 1956.

Caveat Emptor

Finally, let's take a look at another, often overlooked, source of toxins in our lives: consumer products. Unfortunately, the incidents that led to the advent of consumer protection legislation in Europe, Canada and the United States were just as dramatic and deadly as the water and air pollution events just described.

One case in point is radium, the discovery of which in 1898 by Marie Curie and her husband, Pierre, led to its use as an internal medicine in the early 20th century. During that time radium was used to treat cancer, anemia, gout and other ailments. The substance was embraced as a cure-all until women working in clock-dial-painting factories started to glow in the dark.

Pocket watches were commonly used by men, but for those who found themselves fighting in the trenches of Europe during World War I, pocket watches were difficult to manoeuvre and hard to read. The luminous quality of radium made it ideal for use in watches that would glow in the dark. So during the war the U.S. Radium Corporation started to manufacture luminescent wristwatches, which became a huge hit for the men in service as well as the folks back home.

Painting the radium onto the watches was a specialized job. In the 1920s women employed as dial painters would bring their brushes, dipped in radium, up to their lips to form a tip that would help them paint on the numbers. Many of the women came to experience severe dental problems, including necrosis, and ailments such as anemia. In 1927 things came to a head when five women from New Jersey filed a lawsuit against U.S. Radium Corporation for negligence in creating dangerous working conditions. The plaintiffs argued that their ailments were caused by radium that had formed deposits in their bones. All five of the women died from radiation-induced cancer within a few years of the suit being settled in 1928. And luminescent watches across the U.S. were thrown into the trash by anxious consumers.

This was not the first acute toxic poisoning experienced by factory workers in the Western world. In the 19th century the advent of the "Lucifer match," which could be struck anywhere, led to phosphorus poisoning among the people making them—predominantly women and children. White phosphorus was coated on the matches because it increased flammability. But the phosphorus caused symptoms quite similar to those of radium poisoning— including anemia, brittle bones and a horrible ailment known as "phossy jaw." The fumes from phosphorus, it turned out, caused tooth loss, gum swelling and rotting of the jaw bone.

Phossy jaw was diagnosed in the 1860s in England, and by the 1870s efforts were made to prohibit the production of white phosphorus matches. In England a ban came into effect in 1910, and in the United States a law was passed in 1912 to prohibit the manufacture of these matches, though white phosphorus was used in fireworks until the mid-1920s.

Mercury poisoning was yet another example of early chronic industrial disease. While today we might think of mercury in association with fish (particularly tuna), in the 19th century the workers of Danbury, Connecticut, where felt hats were manufactured, were exposed to mercury on a daily basis. The substance was used in a process known as "carroting," in which the fur used to make felt hats was washed in a mercury nitrate solution. The resulting dust and fumes were inhaled by the workers and created symptoms including lethargy, depression, loss of appetite, headaches, ulcerated gums and, in the later stages of poisoning, the shakes. Amazingly, the use of mercury by the felt industry was not banned until 1941.

In the 19th century a variety of chemicals (arsenic, lead, mercury, cyanide, chromium and cadmium) were widely used in pigments for paint and wallpapers. Arsenic helped create a popular green pigment in wallpaper that was so toxic the eminent medical journal *The Lancet* took up a campaign to have it banned. Even William Morris, the founder of the British Arts and Crafts movement, used arsenic green in his line of wallpapers from the mid-1860s onwards and dismissed its poisonous effects.

At about the same time on the other side of the Atlantic, Robert Clark Kedzie, a medical doctor in Michigan who was also a member of the Michigan Board of Health, took up the issue of arsenic wallpaper. Concerned about its toxic effects, Kedzie assembled samples of poisonous wallpaper and had them bound into a book he called *Shadows from the Walls of Death*. He then distributed copies of this toxic volume to libraries throughout Michigan (an original copy of which remains safely, and hermetically, stored at the Special Collections Unit of the Michigan State University Libraries). Although laws against dangerous colorants were common in Europe, industries in America claimed that the concept of public health regulation flew in the face of liberty. As a result American states didn't begin passing laws limiting the amount of arsenic in wallpaper until 1900.[19]

Early work in the U.S. regarding consumer rights resulted in the establishment of the Consumers Union in 1933. Three years later, Arthur Kallet and Frederick Schlink co-authored *100,000,000 Guinea Pigs: Dangers in Everyday Foods, Drugs, and Cosmetics*. This groundbreaking book was "intended not only to report dangerous and largely unsuspected conditions affecting food, drugs and cosmetics, but also, so far as possible, to give the consumer some measure of defence against such conditions."[20]

Around this time deadly events involving consumer products were commonplace. The impetus for enacting the U.S. Food, Drug and Cosmetic Act in 1938, for example, was an incident involving a substance called "elixir sulphanilamide" that was used to treat streptococcal infections. The drug was originally made in tablet form, but demand for a liquid form prompted its development and subsequent shipping prior to toxicity testing. As a result over one hundred people died.[21]

The creation of the first flammability regulations to protect consumers was brought about by a series of deadly fires caused by new ingredients in fabrics in the years following World War II. During the Christmas season of 1951, "torch sweaters" became all the rage. The sweaters were made of brushed rayon and in some circumstances would explode when a spark was dropped on them.[22] Children were also affected. In one case young Michael Blessington was burned to death when his "Gene Autry" cowboy suit caught fire. It turns out that the chaps in the suit were made of flammable rayon.[23] The worst and perhaps most bizarre incident involved a woman who was critically burned when the netted underskirt in her ball gown exploded. The underskirt was made from nitrocellulose (the basis of gunpowder) and ignited in a rather dramatic fashion at a New Year's Eve party.[24]

All of these experiences culminated in the adoption of the U.S. Flammable Fabrics Act in 1953.

Tooth Fairy

If the water dispensed from the water pump in your neighbourhood is killing people or if you want to cut back on the smog from a belching smokestack or if the radioactive paint in your watch is causing people's teeth to fall out, the problem is clear. So is the solution. You can see what's producing the obvious health effect.

And you know that by getting rid of the pump handle or cleaning up the smokestack or changing the paint to a non-toxic alternative, the beneficial effect will be immediate.

But the question that Ken Cook wrestled with in 1998 was what to do when the pollution you're facing is global in its scope, is largely invisible and is—along with a mixture of other chemicals at low levels—affecting human health in a chronic and less obvious manner. "These global pollutants are measured at trace levels, but we now understand that some of those levels can very dramatically affect the human body, hormone systems and the immune system. They can trigger neurological problems and so forth," says Cook. "We've gone from the sort of acute poisoning events of people dying from air pollution to this slow poisoning and links to many chronic diseases that have a chemical component to them."

But as an advocate interested in cleaning up this kind of insidious pollution, how do you render your argument into the "concrete shape" that people "can carry home with them," as Emerson urges in the opening quotation of this chapter?

A couple of examples from recent history point the way.

In 1970 the Environmental Defense Fund took out an ad in the *New York Times* with an edgy and provocative headline: "Is mother's milk fit for human consumption?" This was the ad that lingered in Ken Cook's memory. It was one of the first times that such a personal approach to pollution was used to build public awareness and campaign for change. The Environmental Defense Fund had discovered that levels of DDT were up to seven times higher in human milk than in milk sold in stores. The ad was part of EDF's campaign to ban DDT—a crusade that was ultimately successful in the U.S. in 1972.[25]

IS MOTHER'S MILK FIT FOR HUMAN CONSUMPTION?

Nobody knows. But if it were on the market it could be confiscated by the Food and Drug Administration. Why? **Too much DDT.** We get it from the food we eat. It's in mother's milk, and in the body of virtually every animal on Earth — including man. DDT kills birds and fish, interferes with their reproduction, decimates their populations. It causes cancer in laboratory test animals, and people killed by cancer carry more than twice as much DDT as the rest of us. **Nobody knows for sure what DDT is doing to humans.** But who wants to wait around to find out?

That's what this country is doing. Waiting. There's been a lot of talk, but little action. You heard DDT was banned. It wasn't. Those were just empty headlines. DDT is still being used, despite acceptable alternatives.

Intolerable? Of course. It is also **illegal.** Did you know that? EDF knows it. Two big federal agencies that are supposed to protect us are not doing their job. EDF has taken them to court to see that they do.

EDF goes to court to protect the environment.

The Environmental Defense Fund's 1970 *New York Times* advertisement

While the Environmental Defense Fund used the evocative image of DDT levels in human breastmilk, a decade earlier a group of concerned and innovative citizens applied the concept of body burden testing to what became known as "the Tooth Fairy Survey," an unlikely marrying of dental iconography with antinuclear campaigning.

At the height of the Cold War, in the late 1950s, the U.S. regularly tested atomic bombs above ground. Concerns about nuclear fallout reached a feverish pitch, and one group of concerned citizens hit upon a novel way to advocate for change. While the bomb testing occurred in Nevada, wind patterns carried radioactive elements, including strontium-90—a byproduct of the fission between uranium and plutonium—far beyond the Nevada desert.

Strontium-90 was known to be hazardous, but there was little study of its effects on humans because it was assumed that it would remain trapped, harmless and out of the way, high up in the stratosphere.[26] This assumption was proven incorrect when strontium-90 began returning to earth much more quickly than anyone had anticipated.

It turns out that strontium-90 is similar to calcium in terms of its chemical characteristics. This revelation caused scientists to quickly become concerned about the effects it would have on grazing cattle, and the Atomic Energy Commission was forced to acknowledge, in 1956, that milk was the most significant source of strontium-90 in human food.[27] Scientific papers started to explore the issue of strontium-90 absorption in human bodies, including in teeth, given its ability to bond to tooth and bone.[28]

In St. Louis, Missouri, over a thousand miles and four states away from the nuclear test site, the St. Louis Citizens Committee for Nuclear Information (CNI) was already drawing public attention to the issue of nuclear weapons testing through its speakers' bureau when it noticed the evidence of strontium-90 being deposited in people's teeth. In a brilliant move the CNI decided to further dramatize the risks posed by strontium-90 to future generations through the collection of baby teeth. The goal was ambitious. CNI wasn't interested in only a few baby teeth; the St. Louis Baby Tooth Survey aimed to collect fifty thousand baby teeth for analysis in order to produce a statistically relevant body of knowledge regarding strontium levels in children.[29]

An article in *The Nation* in 1959 noted that "the importance of an immediate collection of baby teeth lies in the fact that teeth now shed by children represent an irreplaceable source of scientific information about the absorption of strontium-90 in the human body. Beginning about ten years ago, strontium-90 from nuclear test fallout began to reach the earth and to contaminate human food. Deciduous teeth now being shed were formed from the minerals present in food eaten by mothers and infants during the period

1948–1953—the first few years of the fallout era—and therefore represent invaluable baseline information with which analysis of later teeth and bones can be compared. Unless a collection of deciduous teeth is started immediately, scientists will lose the chance to learn how much strontium-90 human beings absorbed during the first years of the atomic age."[30]

While slow to catch on in the beginning, the efforts of the CNI gradually picked up steam and captivated the public's imagination. Media attention was substantial, and children from across the U.S. and Canada sent their teeth to the St. Louis "tooth fairy." For their contributions, they were sent a button that read "I gave my tooth to science."

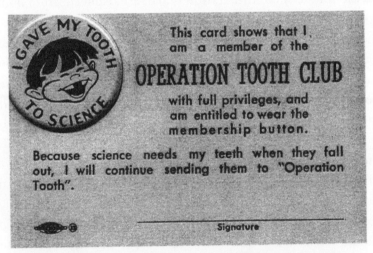

I GAVE MY TOOTH TO SCIENCE

This card shows that I, am a member of the

OPERATION TOOTH CLUB

with full privileges, and am entitled to wear the membership button.

Because science needs my teeth when they fall out, I will continue sending them to "Operation Tooth".

Signature

Button and membership certificate of the Operation Tooth Club.[31] Both were sent to children who donated their teeth to the Committee for Nuclear Information.

The initial results of the survey, released in 1961, demonstrated that the presence of strontium-90 in children was increasing. Teeth from 1951 to 1952 contained strontium-90 at levels of approximately 0.2 microcuries per gram. By the end of 1953, the number had doubled, and by 1954 it had quadrupled. In the end the examination of baby teeth revealed that as a result of increased nuclear testing, levels of strontium-90 had soared by 300 per cent from 1951 to 1955.[32]

The Tooth Fairy Survey exemplified the power of using science in innovative ways to advance public policy goals. Scientists, doctors and dentists collaborated on the survey and worked closely with parents right across the U.S. and Canada. Parents and their children, through donations of teeth, made the effort possible.

The project was very effective. While CNI was not a political organization, the results of the survey were undeniably significant and played an important role in discussions of nuclear fallout. The public was mobilized, and increased pressure was brought to bear on the Kennedy administration to ban atmospheric atom bomb testing. The Test Ban Treaty was signed in 1963, by which point a mind-boggling one hundred and thirty-two thousand baby teeth had been collected.[33] In an address a year later, President Lyndon Johnson specifically mentioned the accumulation of strontium-90 in children's teeth as one of the horrors the Test Ban Treaty was intended to avert.

Human Toxome

Like the Tooth Fairy campaign, body burden testing, since its introduction by EWG in 2001, has taken off.

Around the same time that Ken Cook brought the idea to his colleagues at EWG, European environmentalists also started shifting their attention from talking about chemicals in the environment to actually testing people's bodies. Suddenly, on both sides of the Atlantic, the collection of blood and urine samples was moving the debate in the same direction.

The European effort was initiated by World Wildlife Fund in the U.K. in 2003. The organization tested 155 people as part of a campaign that travelled throughout the country. WWF and Greenpeace Netherlands then conducted further studies on celebrities, members of the public and members of Parliament from a variety of countries in the European Union (among them, a significant number of European environment ministers).

By 2006 the idea of measuring one's body burden was spreading like wildfire. Groups across the U.S.—in Washington State, Connecticut, Massachusetts, Alaska, Maine, Illinois, New York, Oregon, California, Minnesota and Michigan—as well as our own Toxic Nation campaign in Canada, conducted multiple body burden tests on local residents.

All of these data convincingly demonstrated that regardless of age, ethnicity, place of work or residence, everyone is contaminated. Even the most clean living among us are polluted. And even the youngest are vulnerable. Unborn babies were found to have hundreds of chemicals in their little bodies, clearly indicating that toxins are passed on to children not only through breastmilk during nursing, but also through the placenta during pregnancy.

We surveyed the 27 body burden studies conducted by environmental organizations around the world in order to draw some conclusions. Although the number of people tested and the number of chemicals for which they were tested varied from study to study, overall more than 690 people were examined for more than 500 discrete chemicals. The maximum number of chemicals tested for in an individual at any one time was 413. All individuals tested for PCBs and organochlorine pesticides were found to have the chemicals in their bodies. Among these individuals were young people born after PCBs had been banned. Some chemical levels were highly variable. For instance, some people had higher levels of polybrominated diphenyl ethers (PBDEs), phthalates or bisphenol A, while others did not even have detectable amounts. Chemical levels varied among family members and residents of the same towns and cities, indicating that sources of exposure are highly specific and not well understood.

Some trends were obvious. Chemicals such as perfluorinated compounds (PFCs) were found in children at higher levels than in their parents and grandparents, revealing that in some cases, pollution is getting worse. But some good news was also in evidence. For example, chemicals that have been phased out or banned are now at lower levels in young people, indicating that chemical buildup is a solvable problem, says Ken Cook. "When we take action, our blood gets cleaner. We took the lead out of gasoline and blood lead levels in Americans went down—we still had problems, but big progress. We see the same thing with PCBs. We still find it in people that we test, because it's still lingering in the environment, but blood levels by and large have gone down. Blood levels of DDT [banned a few decades ago] have also gone down."

In Europe, body burden testing by environmental organizations was a key component in the campaign to implement the Registration, Evaluation, Authorization and Restriction of Chemical Substances program (REACH), an E.U.-wide initiative to regulate chemicals. On this side of the Atlantic, Canada's recently introduced Chemicals Management Plan (CMP) has been propelled to a great extent by our Toxic Nation testing. And in the United States, the collective efforts of organizations using body burden tests and the leadership provided by the Environmental Working Group have resulted in the introduction of the Kid-Safe Chemical Act in the U.S. Congress. This act "has a provision that says if a chemical occurs in umbilical cord blood it will be deemed unsafe until proven otherwise," says Ken Cook. "We can't fool around when it comes to infants and children. Reversing the burden of proof [which under current law says that a chemical is deemed safe unless proven otherwise] and using the biomonitoring data themselves as a policy instrument: That's the ultimate application of the body burden approach."

Our Toxic Experiment

That brings us to the point of this book. Body burden testing has demonstrated that we're all marinating in chemicals every day. But where exactly are these things coming from? What brands of products are responsible? Can the toxins be avoided in an effective way? And will changes in behaviour—or, preferably, government policy controlling the problem chemicals—result in appreciable improvements in our personal pollution levels?

In the following chapters, we explore seven chemicals that, with the exception of mercury, are more dangerous now than they have ever been. Their production is increasing. And the number of products in which they're used has exploded. At the same time the levels of these chemicals in human bodies are rising.

Our experiments simply mimic what many people would normally do in the course of any given day. They highlight the link between the most ordinary activities and a measurable increase in levels of pollution in our bodies. To carry out the experiments, we (Rick and Bruce) spent one week exposing ourselves to a variety of pollutants. But while we were voluntarily and deliberately exposing ourselves to these substances, thousands of people were unknowingly and involuntarily exposing themselves to the same chemicals.

In order to make changes in our toxin levels more easily detectable, we limited our exposure to the pollutants in the days leading up to the tests. Bruce avoided eating fish for one month before the tests, and Rick tried to steer clear of phthalates, bisphenol A and triclosan for 2 days (48 hours) prior to the tests. We measured any increases or decreases by methodically taking blood and urine samples before and after performing our planned activities.

It seemed easiest to do our testing together. So the bulk of it occurred in Bruce's condo over the better part of two days. We stayed in our "test room" in the condo on 12-hour shifts, which created a pattern not unlike our regular routines, but we were confined to this room. About 10 by 12 1/2 feet, the room was much like any bedroom, TV room or home office in any apartment across North America. Here we experimented on our own chemical loads, exposing ourselves to "everyday chemicals," the ones contained in personal-care products like phthalates and triclosan, the bisphenol A that leaches from baby bottles, the mercury in tuna and off-gassing from carpets.

So Rick showered and washed dishes, he drank coffee in a poly-carbonate cup and ate lunch heated in a container in the micro-wave. Bruce ate tuna and then a little more tuna, and we had the test-room carpet protected with STAINMASTER. There was little more to it. You can see the procedure for yourself in our test schedule, which appears below. Between the planned activities and the donation of blood and urine to the nurse we hired, we caught up on our reading. We watched a lot of CNN. We played Guitar Hero.

By the end of the week, we were freed from the confines of our little test room and back to our regular routines. And our blood and urine samples were sent to Sidney, British Columbia, to be analyzed at Axys Analytical Services, a highly respected laboratory, that does a lot of work for governments and police forces across the continent.

Then, with our polluted bodies, we returned to our usual lives while awaiting the results.

RICK AND BRUCE'S TEST SCHEDULE

Saturday, March 1, 2008

Rick limits exposure to products that contain phthalates, bisphenol A and triclosan.

Sunday, March 2, 2008 (Day 1)

Rick continues to limit exposure to products containing phthalates, bisphenol A and triclosan.

Rick begins the 1st 24-hour urine collection.

1 p.m.   Rick & Bruce meet to have 1st blood samples taken.

2 p.m.   Bruce has 2 tuna sandwiches for lunch.

Monday, March 3, 2008 (Day 2)

9 a.m.   Rick & Bruce arrive at the condo.

9:45 a.m. Bruce drinks Earl Grey tea.

10:15 a.m. Carpet cleaning company arrives to protect/STAIN-MASTER the test-room carpet & couch.

11 a.m.   Rick drinks 1st coffee, brewed in polycarbonate French press.

Rick gets ready for the day (showers, shaves, brushes teeth, etc.).

11:30 a.m. Rick & Bruce settle into the test room.

12:15 p.m. Rick washes hands with antibacterial hand soap.

1 p.m.   Bruce has tuna sandwich & tea for lunch.

1:30 p.m. Rick has chicken noodle soup & canned spaghetti for lunch. (Both were microwaved in Rubbermaid microwavable containers.) Rick also brewed a fresh pot of coffee.

2 p.m.   Rick begins 2nd 24-hour urine collection and takes a urine spot sample.

2:30 p.m. Rick does dishes and washes up, then uses lotion (brushes teeth & washes hands).

3 p.m.   Bruce has a tuna sandwich & tea for a mid-afternoon snack.

3:15 p.m. Rick drinks 2 small (275-mL) cans of Coke.

4:30 p.m. Rick brews fresh coffee and then drinks it.

5:15 p.m. Rick & Bruce have 2nd blood samples taken.

5:45 p.m. Bruce has a trayful of tuna sushi and sashimi.

6:45 p.m. Rick has tuna casserole for dinner.

7 p.m.   Bruce eats a trayful of tuna sashimi, sushi roll and nigiri sushi for dinner, along with a beer or two.

7:15 p.m. Rick washes dishes, washes hands & brushes teeth.

8:15 p.m. Rick moisturizes hands.

9:00 p.m. Rick takes 2nd urine spot sample.

9:30 p.m. Rick & Bruce leave the condo for the night.

Tuesday, March 4, 2008 (Day 3)

10 a.m.   Rick arrives at the condo and makes 1st cup of coffee of the day. He settles into the room.

11 a.m.   Rick brews fresh pot of coffee and plugs in air freshener in the room.

11 a.m.   Bruce arrives at the condo and settles into the room.

11:15 a.m. Rick showers.

11:45 a.m. Rick has canned pineapple for a snack.

1 p.m.   Rick unplugs air freshener and removes it from the room and then makes lunch.

1 p.m.   Bruce has tuna sandwich for lunch.

3 p.m.   Rick takes 3rd urine spot sample.

7 p.m.   Bruce has seared tuna steak for dinner.

9 p.m.   Rick & Bruce leave the condo for the night.

Wednesday, March 5, 2008 (Day 4)

9:30 a.m. Rick & Bruce have final blood samples taken. Rick had additional samples of blood drawn to analyze his PBDE levels.

12 noon   All blood & urine samples shipped to Axys Analytical Services.

Thursday, March 6, 2008

10 a.m.   Blood & urine samples arrive at Axys.

TWO: RUBBER DUCK WARS

[IN WHICH RICK QUESTIONS TOXIC TOYS]

To attract men, I wear a perfume called "New Car Interior."

—RITA RUDNER

THERE ARE A LOT OF LITTLE BOYS running in and out of our house.

My two sons, Zack and Owain, are aged four and one. My sister has two sons. Most of our neighbours have sons. With only a few exceptions, all our good friends from university have had boys. We joke that it must be something in the water of the Haliburton County home, owned by our friends Paul and Irene, where we all gather in the summer.

So I can tell you that as the bearer of a Y chromosome—and the father, uncle and honorary uncle of many others—Dr. Shanna Swan's study on the effects of phthalates on human health scared the beejeebers out of me.

Call me crazy, but I just tend to have that reaction when confronted with compelling evidence that levels of phthalates currently in the environment are very possibly screwing up our children's testicular function. Smaller penis size, incomplete testicular descent and little kids with scrotums that are small and "not distinct from surrounding tissue"[1] are the highlights.

"Demasculinization" is one way that Swan has tried to describe all of the above. A weird but evocative word that's going to stick with me for a while.

The chemical industry has certainly taken Swan seriously, judging from the volleys of criticism they started lobbing her way as soon as her research was published in 2005. While her colleagues in the scientific community call her work seminal, groundbreaking and troubling, industry lobbyists have labelled it flawed and premature and hinted darkly at "data fiddling."[2] As one measure of her impact, Swan remains the only researcher specifically mentioned in the "Media Information Kit" on the chemical industry's phthalates propaganda website. Torpedoing her credibility with journalists would appear to be their job number one.

So what did Swan and her colleagues do to so greatly annoy the who's who of big companies—including BASF, Eastman Chemical Company and ExxonMobil Chemical—that manufacture this difficult-to-pronounce and controversial substance? They did some good scientific detective work, that's what.

Rodents and Humans

In a world where acronyms are overused, the worst one I've ever seen is GLSLRBSWRA. Believe it or not this is a real thing and it stands for Great Lakes–St. Lawrence River Basin Sustainable Water Resources Agreement. One of my colleagues was going to use this in a letter recently until I objected on the grounds that Hell must surely have a special place reserved for those who use acronyms with a double-digit number of letters.

Though they're slightly shorter, I'm about to throw a number of acronyms at you. This is, I'm afraid, unavoidable, given that "phthalates" are actually a class of more than a dozen commonly used chemicals found in numerous household products. Different types of phthalates include diethylhexyl phthalate (DEHP), di-isodecyl phthalate (DIDP), di-isononyl phthalate (DINP) and many others. Global phthalate production is estimated to be in excess of 8.1 billion kilograms a year. Big stuff.

All phthalates look like clear vegetable oil, and it's really their greasiness that makes them useful. Some, like those mentioned above, are commonly used as "plasticizers," keeping substances like vinyl—which would otherwise be hard and brittle—soft and rubbery. In fact, the vast majority of phthalates production—certainly in excess of 60 per cent—is used to plasticize vinyl. Another phthalate, diethyl phthalate (DEP), has a very different purpose: It has become ubiquitous in personal-care products. It helps lubricate other substances in the formula, allows lotions to penetrate and soften the skin and helps the fragrance in scented products last longer. Chances are that if you have smelly air fresheners, toilet bowl cleaners or shampoos (among a multitude of other things) in your home, they contain phthalates.

Because of all these different uses, phthalates permeate the environment—and humans. In the United States, the Centers for Disease Control and Prevention (CDC) have been looking at the blood and urine levels of various pollutants in the bodies of Americans for almost ten years and have found a variety of phthalate products, or "metabolites," in virtually everyone tested.[3] Similar results have been found in Europe.[4]

So the bad news about phthalates is that we all have a bunch of them in us (because there's little reason to believe that results would differ much between industrialized countries).

The good news is that unlike other chemicals discussed in this book (e.g., PFOA in Chapter 3), phthalates break down quickly in the human body and in the environment. If we stopped making them tomorrow, the global contamination would disappear from most places relatively quickly—with the exception of isolated environments like deep sediments in lakes and oceans.

Scientific interest in, and research on, the effects of phthalates has been growing rapidly. A variety of studies on mice and rats have linked fetal exposure with what's come to be called the "phthalate syndrome." It includes a decrease in the distance between the anus and the base of the penis, incomplete testicular descent and a birth defect of the penis called "hypospadias"—a deformity in which the urethra doesn't open at the tip of the penis but rather somewhere along its length. It also increases the risk of testicular cancer in adulthood and of impaired sperm quality.[5]

At the same time scientists studying people have noted a pattern of male abnormalities that they have labelled "testicular dysgenesis syndrome" (TDS). This includes hypospadias, impaired sperm quality, testicular cancer and cryptorchidism (the absence of one or both testes from the scrotum). (While cryptorchidism is common at birth, in most cases the absent testis descends spontaneously by one year of age.) One leading theory for the cause of TDS has been exposure to hormone-disrupting chemicals like phthalates.[6]

The question that Swan, a Professor in the Department of Obstetrics and Gynecology at the University of Rochester in New York, asked is whether these two syndromes could be connected. And through studying pregnant women, mothers and children from several U.S. locations, she found evidence that they are. The higher the level of phthalates in the mother, the more likely that her boy would show signs of phthalate syndrome. As Swan noted in her study, "The present findings, though based on small numbers, provide the first data in humans linking measured levels of prenatal phthalates to outcomes that are consistent with this proposed syndrome."[7]

Importantly, these changes occurred at real-world phthalate levels that have been measured in about one-quarter of women in the United States, and Swan's work highlighted the effects of DEP, a smaller-molecule phthalate that industry has always sworn up and down is completely safe, even while they grudgingly start removing other phthalates from their product lines.[8]

No wonder the chemical companies target Swan so specifically. Disturbing evidence from rodent studies is one thing. But data from real humans? Yikes! This calls for some extra-special denigration of the researchers involved. For Swan's part, she's quite sanguine about all the industry attention: "The industry does its job, which is to protect its products and to attack the science that criticizes their products. My job is to do the best science that I can."

Swan undoubtedly furthered her status as industry *persona non grata* with her co-authorship of another study that looked at phthalates in the urine of 163 infants. Over 80 per cent of the kids in the study had detectable phthalate levels, and infant exposure to phthalate-containing lotion, powder and shampoo (as indicated through surveys of the parents) were significantly associated with increased urinary concentrations of three phthalate metabolites: monoethyl phthalate (MEP), monomethyl phthalate (MMP) and monoisobutyl phthalate (MiBP). The more of these products a mom reported using, the higher her infant's phthalate level.[9] Swan is particularly concerned about this simultaneous exposure to multiple phthalates: "These phthalates add up. If you're exposed to five, six, seven of them at a very low level, together you get a substantial effect. This is what I think is going on in the human studies."

The only bright spot for my household in these otherwise depressing results is that neither diaper cream nor baby wipes (we go through a lot of those) seemed to contribute to the infants' phthalate levels. But as we'll see, personal-care products are not the only things that crank up the levels of phthalates in both adults and kids.

Where Do They Come From?

When I was in high school, my friends and I would say—stealing the line from the movie *The Blues Brothers*—that we were "on a mission from God" if we felt particularly driven to do something. After reading Dr. Swan's study, that was sort of how I felt when I reflected on the implications for my kids. Where were these phthalates coming from? How could I get them out of Zack's and Owain's lives?

One of the most complete summaries of human exposure to phthalates via consumer products was written by Dr. Ted Schettler, the Science Director at a nonprofit research institute in the U.S. called Science and Environmental Health Network. Schettler is a former emergency medicine doctor who decided in the early 90s to dedicate himself full time to the investigation of the links between human health and the environment. "It seemed like a natural extension of what I was doing with individual patients," he explained. When I reached him at his home in Ann Arbor, Michigan, I asked him to describe where kids might become exposed to phthalates.

"Not an easy answer," was his response. "The phthalates in your kids' lives are going to be coming from many products that are in your house and in their daycare and school. The phthalates are going to migrate out of those products, either onto your kid's hands or onto furniture or the floor, and they're going to contaminate the dust in your children's room and your living room. They're going to get phthalates through the food they're eating. And they're also going to get them through the products that are put on their skin."

In addition to many vinyl products often found in the home—toys, shower curtains, raincoats and the like—phthalates are found in things my boys don't currently have any exposure to, such as building materials, blood bags, intravenous fluid bags and infusion sets, and other medical devices. Because the phthalates are not tightly bound to the plastic or vinyl in these items, they easily leach out of them. The interiors of most new cars, referred to in the introductory quote to this chapter, are lousy with phthalates.

"Will the phthalates exposure of my boys be similar to mine?" I asked.

"To a certain extent," Schettler answered. "The child is going to encounter the same environment as an adult but in a different way. They're going to be playing and moving around in it in a different way and putting their fingers in their mouths much more frequently than you are. They're going to be more intimately in contact with their physical environment than adults are, and this will be reflected in their level of exposure."

In other words, by virtue of being closer to the dust bunnies, licking their fingers relentlessly and chewing on phthalate-containing items that they shouldn't be putting in their mouths, my kids are sucking in more of this stuff than I am.

Recent studies support this conclusion. A report that looked at several phthalates in children from California found levels in all the children examined to be higher on average than levels discovered in a similar study of adults. The researchers concluded that "DBP, BBP [dibutyl phthalate and benzyl butyl phthalate, plasticizers both commonly used in vinyl], and DEHP exposure on a body weight basis may be at least twice as high for these children compared to the adults in [the initial study]."[10] In addition, the major 2003 and 2005 CDC reports mentioned earlier in this chapter found levels of a number of phthalates to be higher in children and in women of reproductive age.

This greater exposure for kids is a serious problem in its own right but is compounded by the fact that kids are at greater risk of harm from pollution than adults. For example, rising rates of childhood asthma have reached epidemic proportions and are linked to air pollution in urban areas.[11] In 2000 the Children's Health Study, led by the University of Southern California, revealed that common air pollutants slow children's lung development. The study spans ten years of monitoring among three thousand students in over one dozen communities in California. "The researchers showed that as children grow up, those who breathe smoggier air tend to lag in lung function growth behind children who breathe cleaner air. Children with decreased lung function may be more susceptible to respiratory disease and may be more likely to have chronic respiratory problems as adults."[12]

The same is true for other, less visible types of pollution. The immature bodies of growing children lack certain detoxification mechanisms and are more prone to the damaging effects of substances like phthalates.[13] Their cells are dividing at an amazing rate. Their organs are developing. During these incredible periods of growth, children are particularly susceptible to damage or disruption by chemical agents.[14] Also important to note is that children's exposure begins at conception as chemicals, including phthalates, cross the placenta in a pregnant woman's body and can do damage to the fetus.[15] The National Academy of Sciences in the U.S. has estimated that 25 per cent of developmental and neurological problems in children could be caused by environmental pollution combined with genetic factors. It cites the increase in low-birthweight births, premature births, atrial septal defects, genito-urinary defects, attention deficit hyperactivity disorder and autism.[16]

Though all of the above is somewhat scary, I took heart from something Dr. Schettler said: phthalates are flushed from the human body—even in kids—very quickly. So if I could remove some of the sources of phthalates from our house, the resulting decrease in pollution in the bodies of my family members would be almost immediate.

In My Kids' Closet

Armed with this information I began sifting through my boys' typical day.

Starting with breakfast I made my way to the kitchen. According to Ted Schettler, phthalate contamination of food is variable but widespread: "Food is probably a major avenue of exposure to DEHP for most people. You can find DEHP in soils and sediments and sludge that are going into rivers. There's a general environmental contamination, so many foods coming from the land will be contaminated. During the food preparation and packaging, phthalates can get in as well." Schettler says that as far as he knows, phthalates are not taken up by vegetables, but because they are fat soluble, they'll get into meat, dairy and processed foods.

Whether or not phthalates are in food packaging is a murky area. As Schettler puts it the industry "asserts with great authority" that phthalates are not in things like cling wrap in the United States. However, I found a couple of studies revealing that phthalates exist in plastic wrap in other parts of the world. Given that labelling is not required on any of these products in Canada, the status of these items in this country is confusing.

It is certainly clear that some food processing methods add

phthalates to the finished products. Schettler mentioned a study from Japan, showing that phthalates in vinyl gloves worn by people handling and packaging food was one source of exposure for consumers. With surveys showing the highest phthalate levels in fatty foods, such as dairy (including infant formulas), fish, meat and oils, Schettler told me he strongly suspected that in addition to entering milk because of their presence in the diet of dairy animals, phthalates in milk products can also be traced back to the flexible vinyl tubing used in dairy barns to drain milk from milking machines into collection vessels.

Truth be told, the source of phthalates in many foods is a bit of a mystery. The same goes for what Jen and I as parents could really do about it. We buy organic food as much as possible. We feed our kids fresh foods whenever we can and not too much red meat. We minimize plastics in our kitchen and certainly never use plastic containers to microwave food. Though these habits are worthwhile for other reasons (because plastics leach stuff into food when heated, period), it's not clear that any of them would have an impact on the level of phthalates in our kids. With two hungry, growing boys, eliminating milk and favourite dairy products like yogourt and Fudgsicles just wasn't an option.

A bit stumped on the food question, I moved on to Zack and Owain's basement toy box.

Here, the sources of phthalates were much more obvious. Many of their action figures, balls, hockey pucks and other toys were made—in whole or in part—from rubbery materials that might well contain phthalates. Though he's certainly not encouraged to do so, Owain chews on many things—toys of all descriptions, sticks, bugs, my ear—especially when he's teething. I wondered about the extent to which these toys leach phthalates into his mouth and how much they contribute to the phthalate levels in our house dust as they disintegrate (which, after much rough handling, they very often do)—though I hasten to add that we're opposed to dust bunnies in our house and vacuum frequently.

The American Academy of Pediatrics has confirmed that ingestion of phthalates can occur when children mouth, suck or chew on phthalate-containing toys or other objects. Though the academy acknowledges that these sources of exposure are difficult to quantify directly, it noted that this type of "non-dietary ingestion can be expected to increase total exposure by an order of magnitude or more" and that "in the United States and Canada, this uncertainty in predicting exposure levels, especially in very young children and infants, has led to the removal of all phthalates from infant bottle nipples, pacifiers, teethers and infant toys intended for mouthing. DINP has been substituted for the more toxic DEHP in many other toys intended for older children."[17]

I made my way upstairs to the kids' bedrooms. Here again, I counted many rubbery toys that might be possible sources. I finished up in the bathroom, staring at the net full of rubber bath toys suspended under the shower head. I also noticed for the first time that the house we had just moved into (and haven't had a chance to fully redecorate) had a fabric, rather than a vinyl, shower curtain. Other than possibly some of the kid toys I'd inventoried, our house was relatively free of vinyl and the many phthalates vinyl contains.

I finished off my tour by examining the kids' soap and shampoo. "Fragrance" it said on the back of one; "*Parfum*," on the other. Because of nonexistent labelling requirements in North America (except, for some chemicals, in California), "phthalates" are almost never listed as an ingredient on products that contain them. "Fragrance" and "*Parfum*" are often the code words indicating some phthalate content.

And so, after investing about an hour in this phthalate treasure hunt, my conclusion was that my kids' toys and their personal-care products were the main sources of phthalates that I could control.

Santa's Evil Elves

I have to admit that I find the role of toy companies in exposing kids to toxins very surprising. It's a bit like realizing that Santa's elves long ago started consorting with Darth Vader and were "turned to the dark side" without anyone noticing. Like many parents I clued into the fact that toy safety standards have, shall we say, slipped in recent years only when Zack's "Thomas the Tank Engine" was one of 1.5 million little trains that were recalled in June 2007. We'd had it for a few years, both Zack and Owain had regularly handled it and put it in their mouths and I wasn't too pleased to find out that its chipped paint was actually full of lead. We dutifully sent the red engine back to the manufacturer and received a replacement a couple of months later. We also received a "gift" from the company for our troubles: a little yellow engine.

Bizarrely, this gift was then recalled in September because it, too, was full of lead. But this sort of surprise was run of the mill in the summer of 2007, when it seemed as if, every day, there was another toy atrocity. Aqua Dots leaching the hallucinogenic date rape drug, GHB. Barbies and Sesame Street dolls painted with lead. Asbestos in a best-selling board game. It went on and on and on. Forty-five million individual toys recalled by dozens of manufacturers, because of a variety of noxious ingredients and serious defects. At one point Mattel, the world's largest toy-maker, had to deliver a personal apology to the Chinese government for harming the reputation of Chinese manufacturers. When the recall avalanche began, Mattel had originally tried to blame lax Chinese safety standards, when in fact it turned out that one of the major problems was a design mistake made by Mattel itself.

Clearly, the toy industry on some very basic and systemic level had lost track of what was being put into its products and by late 2007 still did not have a handle on the problem. Despite the recalls, testing of 1,200 toys by the Washington Toxics Coalition, the Michigan-based Ecology Center and others found that more than a third of the toys still contained lead and nearly 50 per cent were made of vinyl and contained phthalates. The testing also revealed that vinyl toys were more likely to contain toxic metals such as lead and cadmium.

Though embarrassed and somewhat contrite at the revelations of widespread contamination from substances like lead, chromium, bromine and chlorine, the toy industry remained unapologetic about the phthalates they were continuing to include in all manner of products. In fact, the Toy Industry Association (TIA) in the U.S. lobbied aggressively against various measures to ban phthalates in toys in California. Without a hint of irony, TIA's Vice President for Standards and Regulatory Affairs Joan Lawrence said in a statement following the defeat of the first California attempt in 2006: "Today's decision by the California State Assembly is an important victory for the toy industry. While toy safety has always been the top priority for the toy industry, the bill's success would have set a dangerous precedent for legislation based on fear and allegation while ignoring science—that could be used to restrict all types of products and materials. TIA is committed to fighting all of the copycat legislation that is currently being proposed in other states."[18]

The toy industry's attitude toward phthalates, despite their spectacular safety failings, didn't exactly fill me with confidence. In fact, the industry's continuing protestations of innocence reminded me of an old Dan Aykroyd *Saturday Night Live* skit that I saw recently. (My friend Mike is an *SNL* aficionado and has many of the old episodes on DVD. A visit to his house in Ottawa usually comes complete with some quality viewing of classic 1970s comedy.)

In this skit Aykroyd plays a toy executive called Irwin Mainway, who is aggressively questioned by a consumer reporter (played by Candice Bergen) about the safety of his toys. Even when confronted with his dodgy products, which have names like Mr. Skin-Grafter, General Tron's Secret Police Confession Kit and Johnny Switchblade (which springs two sharp knives from its arms), nothing will shake Mainway from his message that his toys are safe. At one point the reporter asks about kids hurting themselves with Bag O'Glass (which retails for $1.98), and Mainway yells: "No! Look, we put a label on every bag that says, 'Kid! Be careful—broken glass!' I mean, we sell a lot of products in the 'Bag O'' line . . . like

Bag O'Glass, Bag O'Nails, Bag O'Bugs, Bag O'Vipers, Bag O'Sulfuric Acid. They're decent toys, you know what I mean?"[19]

Listening to toy companies talk about phthalates in their products is a bit like that: "They're safe! Really they are!"[20]

With visions of Irwin Mainway in my head, I set out to plumb the depths of the phthalates in Zack and Owain's toy box. I decided the direct approach was best, so I chose a few toys at random and sent them away for lab analysis (at STAT Analysis Corporation in Chicago).

Here are the toys as they were returned to us following the testing at STAT. The holes are from the samples drawn for the tests.

The toys from Zack and Owain's toy box that we sent to STAT Analysis to be tested

The results were every bit as worrisome as I'd feared. With the exception of a Baby Einstein bath book, every one of the other toys had phthalate levels above o.1 per cent, the allowable maximum set by European legislation. With the exception of the Kidizoom Camera, which is made for older kids (but which Owain regularly mouths nonetheless), all the toys would likely be prohibited by European and now U.S. legislation, which focuses on banning phthalates in products manufactured for children under the age of three.

One small, chewable, red, white and blue rubber ball—that both Zack and Owain have put in their mouths repeatedly over the years—had incredibly high levels of DEHP, a particular phthalate that even industry sources admit is of concern.[21]

Table 1. Phthalate levels in a random sampling of Zack and Owain's toys

Name of Toy	Manufacturer	Total Phthalate Content (%)	Bis(2-ethylhexyl) DEHP (%)	Diisodecyl DIDP (%)	Diisononyl DINP (%)
Kidizoom Camera; handles	Vtech	0.9	ND	0.9	ND
Rubber Duck; large yellow	Munchkin	0.67	.05	ND	.66
Rubber Duck; small yellow	Unknown	27.0	.051	ND	27
Ball; red, white & blue	Unknown	43.7	41	ND	2.7
Frog; green	Unknown	1.3	ND	ND	1.3
Bathtime book	Baby Einstein	ND	ND	ND	ND

ND = not detected

With the December 2008 pre-Christmas shopping rush in full swing, I phoned the Canadian Toy Testing Council and the Canadian Toy Association with these test results. The conversations were not exactly satisfying.

The Toy Testing Council, which actually ranked the Kidizoom Camera as its "Toy of the Year" for 2008, seemed barely aware of what phthalates were let alone that the toy they so heartily recommended for every parent's Christmas list owed its pleasant rubberiness to DIDP. When I asked Harold Chizick of the industry's Toy Association why it was acceptable that toys with phthalate levels rendering them illegal to sell in the E.U. or the U.S. would

wind up being peddled in Canada, he downplayed the scal
problem. Because many of his association's members are in
tional companies, they are increasingly being guided by the j ...-
dictions with the most stringent regulations, he explained.
Translation: The lack of any phthalate regulation in Canada means
that whatever protections are afforded Canadian kids are being
accomplished by foreign governments.

Because of decisions of the E.U. and the U.S., the industry's
biggest players are now moving away from phthalates, Chizick
said. He warned that many toys for sale in "dollar stores" or other
such discount outlets are not manufactured or imported by major
toy companies and that their ingredients are subject to less stringent oversight. And what he called the "sixty-four-thousand-
dollar question" is how to deal with the legacy of toxic toys in toy
boxes across the globe. Once toys are manufactured with toxic
ingredients, even when governments change regulations to
improve safety standards for children, most parents never hear
of this new information. I'm sure the young families on my street
are similar to those the world over: Toys are passed from one
kid to another within families and between neighbours. Even if
entirely eliminated in newly manufactured toys tomorrow,
phthalates will live for years to come in the toys of millions of
Canadian homes.

My last question for Chizick was whether, in order to qualify for
his association's annual "Hot Toys for the Holidays" ranking, manufacturers are required to provide documentation indicating that
their products are clean of lead, phthalates or other toxins, such as
bisphenol A. There was a long pause on the other end of the line.
"No," he said. "Self-regulation is hard as an industry association."

Caveat emptor, then. How much do you want to bet that more
than a few kids received cheerfully painted, phthalate-ridden toys
as stocking stuffers?

The Sweet Smell of Pollution

The other big question I wanted to answer was whether my kids' exposure to phthalate-containing personal-care products was significant or not. How best to do this?

There was no way I was going to test their blood or urine. They're far too young for that. And the existing studies about the effects of personal-care products on phthalate levels didn't use name-brand, off-the-shelf products. So the next best thing, in order to get a sense of just how easy it is for the human body to absorb these substances, was to experiment with phthalates on myself.

I phoned Dr. Susan Duty at the Harvard School of Public Health for some advice on how to put together the experiment. Duty was the primary author of a very interesting 2005 study that looked at levels of phthalates in the urine of over four hundred men and correlated the data with the kind of personal-care products they reported using. She found a very clear and striking relationship: The more products the men used, the higher their urinary concentrations of MEP. Men who used cologne and aftershave had higher levels than those who did not. Those who used cologne, aftershave, hair gel and deodorant had higher levels still, and so on.

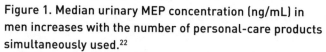

Figure 1. Median urinary MEP concentration (ng/mL) in men increases with the number of personal-care products simultaneously used.[22]

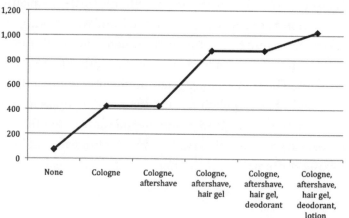

Like most of the scientists we contacted for advice on our experimentation, Duty was intrigued. "If you want to measure increased phthalate levels, you probably want to start by trying to depress the amount of the chemical in you," she advised. "The half-life of phthalates in people is short. About 12 hours. So you want to minimize your contact with sources of phthalates for about a day to get your levels down."

"No problem," I said.

"At the end of this initial period, you should start collecting your urine for 24 hours. You should be careful to avoid phthalates in personal-care products and food that's been contacted by phthalates. Try to choose unprocessed over processed foods."

Ouch. That's going to eliminate my favourite pretzels and jalapeño chips, I thought.

After chatting through some examples of personal-care products that may contain elevated levels of phthalates, Duty finished by telling me, "As you're exposing yourself you have to collect your urine for this entire 24-hour period as well," which gave me visions of my refrigerator crammed with urine bottles. But I thought, *Anything in aid of science.*

And so my shopping adventures began.

Plastic Free

Because I was simultaneously trying to "detox" from phthalates and bisphenol A (Chapter 8), which, between the two of them, are found in a bewildering array of plastics, I set out to go the whole nine yards. To shoot for the moon. To go two whole days without eating anything that had come in contact with plastic.

This is a lot harder than it sounds. Try it. I dare you.

I did this because I wanted to see if I could. But also for a practical reason. Despite the chemical industry's protestations that phthalates aren't in food packaging (at least in the U.S.), I had no way of validating this claim. Just avoiding plastic altogether seemed easier than trying to puzzle over the specific plastic type of every food container I was interested in.

I sort of knew it already, but once you start carefully keeping track, it really hits you: plastic has taken over our lives. All my favourite snacks come swaddled in the stuff. Fruits and veggies from your typical grocery store are always bagged in plastic, either by the grower (seemingly an increasing trend with "prewashed" greens and the like) or stuffed in plastic bags in the produce aisle. And as I went through our kitchen, I started to realize that virtually everything (with a few notable exceptions, such as my favourite Cajun "Butt Burner" hot sauce) are covered in plastic. Mayonnaise (when did Hellmann's change from glass to plastic?), juice, milk, store-bought bread, sauces, cookies, yogurt, etc., etc.

Rotisserie chickens from the grocery store now come in sturdy little plastic boxes. This is a change from my childhood days, when according to our weekly tradition, my father, my sister and I picked up a whole cooked chicken in a foil-lined paper bag at Steinberg's and took it over to my grandmother's apartment to watch that beautiful 1970s Friday-night trifecta: *Donny and Marie, The Love Boat* and *Fantasy Island.*

You name it. Even things that look like they're packed in cardboard, such as breakfast cereal, pasta and Goldfish crackers, are actually encased in plastic pouches on the inside of the box.

For my two days of freedom from plastic, I relied on fresh foods that I purchased at the St. Lawrence Market (the largest farmers' market in Toronto) with Jen and the kids on Saturday morning. Going to the market is a family routine for us. So as Zack and Owain jumped up and down in front of their favourite busker, I picked up my groceries: fresh bagels and bread (bagged in paper), organic soup in a glass bottle, organic fruits and veggies and pasta (packed in cardboard) with pesto (in a glass bottle). This is what I lived on for two days. With the exception of having to demur when offered salty, crunchy snacks from plastic bags at a party that Saturday evening, it wasn't too bad.

Smelling Purty

The other shopping expedition I went on was for the personal-care products I would use to try to crank up my phthalate levels. In order to put this unusual list together, we consulted two helpful sources: the Environmental Working Group's Skin Deep database (www.cosmeticdatabase.com), which provided plenty of good information on the content of various products, and "Not Too Pretty—Phthalates, Beauty Products & the FDA," a 2002 study that tested 72 name-brand, off-the-shelf beauty products for their phthalate content.

Most people, I'm sure, use these publications to *avoid* phthalates. We used them to seek those chemicals out.

We purchased all the products through various Toronto retailers. None of these products listed "phthalates" as an ingredient, though all listed "Fragrance" or "*Parfum.*" And fragrant they certainly were. We stored all our purchases in a cardboard box in a corner of Bruce's condo. The combined smell of the various bottles was quite something—a sickly sweet combination of roses and pine needles.

Table 2. Rick's phthalate shopping list

Hair: Pantene Pro-V sheer volume shampoo & conditioner Pantene body builder mousse TRESemmé European freeze hold hairspray
Shaving: Gillette deep cleansing shave gel
Other Toiletries: Calvin Klein Eternity for men Right Guard sport regular deodorant Jergens original scent lotion
Kitchen: Dawn Ultra Concentrated liquid/antibacterial hand soap – apple blossom scent
Test room: Glade Plug-in Scented Oil – morning walk scent (plugged in for 2 hours on Tuesday)

Because for many years I've tried to buy unscented toiletries, I found the aroma of all the products I was suddenly applying to myself quite annoying. But there are many people in the world who use combinations of these or similar products every day. After all, Coco Chanel, that maven of 20th-century fashion, is reputed to have once said that "a woman who doesn't wear perfume has no future."

The experiment itself was straightforward (see our schedule at the end of Chapter 1). From Friday night through Sunday morning, I limited my phthalate exposure in every way I could think of. I ate the fresh food described above after fasting on Saturday and Sunday. I didn't shower. I avoided anything with heavy scents, including all personal-care products.

From Sunday morning through Monday morning, I continued this regime and collected 24 hours' worth of urine on Monday at

2 p.m. From Monday at 2 p.m. to Tuesday at 3 p.m., I collected the second 24 hours' worth of urine. During this period I showered, shaved and used toiletries and cleaning products in the same way I would have at home. Except in this case the products I was using were those listed above.

After our project coordinator, Sarah, packed up the litres of my urine (very, very carefully) and sundry blood samples and sent them to Axys, we waited. It was over a month before the results were returned. And we really had no idea what to expect. Had the experiment worked?

Results

It worked all right. I was actually shocked at the results.

Of the six phthalates for which we tested, five were present at detectable levels before and after I lathered myself in all the smelly products. Levels of monoethyl phthalate (MEP)—the metabolite of DEP (diethyl phthalate)—were a lot higher in the "Before" measurement than any other phthalate. This just makes sense. According to the "Not So Pretty" testing of off-the-shelf personal-care products, DEP was present in 71 per cent of the products, and DBP, BBP and DEHP (dibutyl phthalate, benzyl butyl phthalate and diethylhexyl phthalate, respectively) were found in fewer than 10 per cent.

The really dramatic result was that as a result of my product use, my MEP levels—one of the chemicals that Shanna Swan had connected with male reproductive problems—went through the roof, from 64 to 1,410 nanograms per millilitre (ng/mL). My MEOHP and MEHHP levels (metabolites of DEHP) declined ever so slightly (from 19 to 10 ng/mL and from 26 to 12 ng/mL, respectively). Interestingly, this increase in MEP and decrease in other metabolites had been observed by Susan Duty and colleagues in their study. Her speculation was that perhaps other ingredients in the products acted to deter the absorption of DBP, BBP and DEHP. It could also be that urinary levels of these metabolites reflect exposure to things other than personal-care products.

Figure 2. Levels of different phthalate metabolites in
Rick's urine before and after exposure to phthalate-containing
personal-care products. DEP is the most common phthalate
in these products, and it converts to the MEP metabolite in
our bodies.

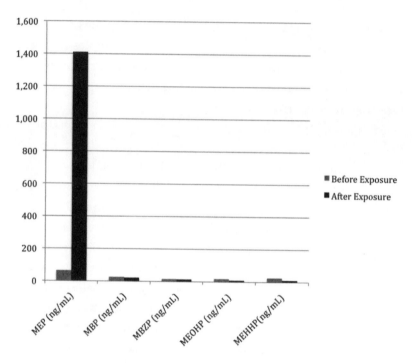

Another interesting aspect of my results was that even after having
that initial "detox" period in which I tried to eliminate all phthalates, I was unable to do so. Five of the six phthalates were there in
measurable levels. Where did they come from? Well, the food I
ate—wholesome as it was—likely contained some phthalates. I also
wondered about a couple of odiferous air fresheners I had run into
on Saturday.[23] One was in a corridor at the St. Lawrence Market;
the other was in the bathroom at our local food store (Zack had to
make an emergency pit stop).

So even after mounting a concerted effort, all I could do was
reduce my phthalate levels, not eliminate them, from my body.

And my little experiment showed how amazingly easy it is to dramatically crank up levels of MEP after a simple change in toiletries for two days.

Who knew that conditioning your hair could be hazardous to your health? And every time I used similarly smelly products on my kids, I was achieving a similar result in their little bodies.

Those Rubber Ducks Rear Their Heads

Sometimes good ideas are infectious. Certainly, as an environmentalist, you always hope so. If you toil away in your city or province or state or country for a great change in government policy, you want to think that it's exportable—that those truly groundbreaking achievements for the protection of the environment or human health will be emulated by other jurisdictions because they're so darn good.

Often this doesn't happen. Rather than learning from each other and adopting successful models, different jurisdictions often end up rehashing the same old debates time and time again. Social change is often slow. Women in New Zealand gained the right to vote in 1893, but it was nearly a hundred years later, in 1990, before all European women could finally say the same. That was when the Federal Supreme Court of Switzerland forced the Canton of Appenzell Innerrhoden to change its knuckle-dragging ways.

What happened with phthalates in kids' toys, however, is a textbook example of what can be achieved in short order if good ideas are spread around.

It wasn't all rosy in the beginning. For a while, in fact, it looked as though Europe and the U.S. were on a collision course. In both places the discussion about what to do about phthalates in children's toys began in 1998. Europe immediately started to move. The next year, on the heels of a variety of studies linking phthalates with human health problems, the E.U. proposed an emergency ban of six phthalates in toys likely to be placed in the mouths of kids under three years of age. The ban, though temporary, was significant. It was the first time the E.U. had enacted an immediate prohibition under its new General Product Safety Directive.

After many twists and turns and millions of euros of chemical industry lobbying later, the action was made permanent in 2005. DEHP (diethylhexyl phthalate), DBP (dibutyl phthalate) and BBP (benzyl butyl phthalate) were banned outright, and DINP (diisononyl phthalate), DIDP (diisodecyl phthalate) and DNOP (di-n-octyl phthalate) were prohibited in toys and childcare articles designed to be sucked or chewed by children under three years old.

In the United States events took a very different course. In 1998, 12 consumer and environmental groups filed a petition urging the U.S. Consumer Product Safety Commission (CPSC) to ban toys containing phthalates intended for children five years of age and under. The CPSC responded by calling for further study to assess phthalate toxicity to humans. Review led to review, process was heaped on more process.

Though the CPSC eventually requested that the toy industry voluntarily remove phthalates from "mouthing toys" and many companies consented (some even removing them from their entire product lines), no recalls ever occurred. Toys remained on store shelves and in family toy boxes throughout the nation. Toy companies continued to claim that they were eliminating phthalates from production because of "consumer concern," not because of any danger from phthalates themselves.

There was another weakness with this voluntary U.S. action as compared to the European regulatory approach. Toy companies and the CPSC used a narrow definition of "mouthing toys," which included only bottle nipples, teethers and rattles. Other toys that children chew on—like rubber bath toys—remained on the market. As any parent can attest, kids will chew on many things if given a chance. My son Owain chews on our cat, for goodness' sake, and despite getting a mouthful of fur is back a few days later for more.

Five years after the initial petition, the CPSC ruled that "there is no demonstrated health risk posed by PVC [polyvinyl chloride] toys or other products intended for children five years of age and under and thus, no justification for either banning PVC use in toys and other products intended for children five years of age and under or for issuing a national advisory on the health risks associated with soft plastic toys."[24] Case closed.

Industry and its allies were positively gleeful. Marian Stanley, for many years the chemical industry's leading spokesperson extolling the virtues of phthalates, issued a press release saying the CPSC decision meant "the great vinyl toy scare is history."[25] Steven Milloy, an adjunct scholar with the conservative Cato Institute, wrote a scathing denunciation of the CPSC petitioners for Fox News entitled "Vinyl Toys Are Just Ducky."[26]

And so the Rubber Duck Wars were joined. With its non-decision at the federal level, the CPSC ensured that the fight would be waged elsewhere. As one of the most charismatic phthalates sources around, the yellow icon, beloved by Sesame Street alumni everywhere, took centre stage in the ongoing U.S. phthalates debate.

From Brussels to San Francisco

In most conflicts throughout the centuries, both sides claimed that God was "on their side" and invoked the divine in defence of their efforts. So it has also been with phthalates and rubber ducks.

Environmentalists have placed this image at the centre of their campaign to highlight the noxious substances that are found in this symbol of childhood innocence.

One hundred smiling rubber ducks and toddlers put pressure on Congress to ban phthalates in toys. The rally was organized by Ann Arbor Ecology Center, Clean Water Action and the Breast Cancer Fund in Ann Arbor, Michigan, June 2008.

Phthalate proponents have positioned themselves as rubber duck defenders, implying that environmentalists want to take away the favourite toys of America's children.

Button distributed by Consumers for Competitive Choice in opposition to a federal ban on phthalates in toys, summer 2008. Note that the duck is sporting a label saying "CPSC [Consumer Product Safety Commission] Approved."

As has repeatedly happened over the past few years, when progress stalls at the federal level in the U.S., advocates for improved environmental and health protection turn to California (which, if it were a nation, would have the tenth-largest economy in the world) to set the pace. Luckily, California environmentalists and public health advocates were up to the challenge.

Janet Nudelman, the Director of Program and Policy for the San Francisco–based Breast Cancer Fund, has been front and centre in the California debate. Nudelman and her organization are unusual among cancer organizations in that they focus on identifying and eliminating the environmental links to the disease. "Only one in 10 breast cancers can be explained by a genetic history for the disease," she explained. "A growing body of scientific evidence is linking environmental toxins to increasing rates of breast cancer and other diseases." On a personal level Nudelman is convinced that an increasing number of people understand this connection: "I don't think I'm the only person of my generation who's looking around, dumbfounded by how many people they know with cancer, and asking themselves: Where is this cancer coming from? Why do so many of my friends have cancer? Will I get cancer?"

The Breast Cancer Fund's interest in phthalates was sparked by increasing evidence that hormonally active chemicals could be, in part, responsible for women reaching puberty at an earlier and earlier age. The trend is clear and worrisome as Nudelman explained: "Black girls are developing breasts between eight and nine years of age when it used to be 10. And white girls are developing breasts at just under 10 when it used to be 11.5. And that's a change that we've seen in just 30 years. As breast cancer advocates, the reason we care about this phenomenon is that the earlier girls experience puberty, the earlier their bodies start to be exposed to estrogen. A women's risk of breast cancer is directly linked to her lifetime exposure to natural and artificial estrogens."

The Breast Cancer Fund believes that women are being dosed with higher and higher levels of artificial estrogens from things like consumer products. The fund is particularly concerned about phthalates because of recent studies that have found elevated phthalate levels in girls suffering from premature breast development.[27] With this concern in mind, the fund girded itself for a fight.

The first round of the phthalates debate in California ended badly. In January 2006 an attempt to institute a European-style ban on phthalates and bisphenol A failed in the state legislature. Now stymied at both the federal and the state levels, advocates like Nudelman decided to take the fight to the only, if unlikely, government left. The municipal level.

Under the leadership of Fiona Ma, a young and energetic member of San Francisco's Board of Supervisors, the Breast Cancer Fund and its allies started working to convince the city to become the first phthalate- and BPA-free zone in North America. This was not without precedent, explained Jared Blumenfeld, Director of San Francisco's Department of the Environment. Since the 1990s San Francisco had blazed a trail of new environmental initiatives. "With respect to chemicals from pesticides to phthalates, we recognized that we shouldn't be doing this. It's crazy," he said. "The local government shouldn't have to regulate these things. And yet the federal government hasn't revamped its toxics laws since the 1970s. It's doing nothing. Absolutely nothing. . . . It's like the U.S. is in the Stone Ages when it comes to caring about its citizens." As a consequence Blumenfeld said, "Laws have started to emanate up from the grassroots. Once action happens in a few cities, it's taken up by the California legislature. And then California is big enough that it can affect the national debate."

And that's exactly what happened with phthalates. With Supervisor Ma's careful shepherding and the strong and vocal support of public health advocates, the city's Board of Supervisors voted unanimously for a ban on phthalates and BPA in certain products in June 2006. Jared Blumenfeld gave some of the credit for the quick progress of the debate and the unanimous vote to the "unyielding arrogance" of the chemical industry itself. "The more sabre-rattling the chemical industry did with the Mayor and Board of Supervisors the more likely they were to vote against the industry. It really backfired on them." The industry threatened lawsuits repeatedly, he said. "At one point I was in a meeting with 16 or 17 lawyers. If nothing else during that period we achieved full employment for their lobbyists! Their goal was to kill the initiative in San Francisco so they could hold the defeat up as an example." Blumenfeld's assessment was blunt: "These guys are the new tobacco lobby. The same people. The same politics."

Following hard on the heels of the city's decision to proceed with the ban, the lawsuits did indeed start flying. First, the American Chemistry Council, the California Retailers Association, the California Grocers Association and the Juvenile Products Manufacturers Association sued. Then the manufacturers of phthalates, the California Chamber of Commerce and the Toy Industry Association piled on with their own legal action. The basis of these suits was a jurisdictional argument. The industry groups alleged that San Francisco simply lacked the authority to ban phthalates and BPA in products within the city.

Determined to figure out a way to proceed in the face of these challenges and to deny the chemical industry the ability to say they had defeated San Francisco, the city began to bob and weave. First, it delayed implementation. The ban was slated to begin on December 1, 2006, before the Christmas season, but the city told businesses it would wait until after the holidays to begin enforcement. Then the Board of Supervisors began to debate amendments. Instead of banning certain phthalates, under the changed legislation labs would be hired to test specific products over the next couple of years. If these products had certain levels of phthalates, sale of those specific toys could be punishable. The fine for the first offence would be one hundred dollars. References to bisphenol A were removed from the legislation entirely.

In response to the delays and amendments, the lawsuits were dropped.

From Sacramento to D.C.

While all this was playing out at a local level, Fiona Ma had been elected to the California State Assembly. She immediately set to work introducing another state-wide phthalates ban, patterned after San Francisco's initiative: the "California Toxic Toys" bill. As Jared Blumenfeld said, "The chemical industry lobby machine then just decamped to Sacramento."

Ma hit the ground running. "Given that I was a freshman in the Assembly I don't think the chemical industry took me seriously at first," she said. "But I worked it. As industry lobbyists were walking out of my colleagues' offices, I was walking in. We had a press conference with a 30-foot-tall, inflatable rubber duck. We got Steven Spielberg and other prominent Californians to write the Governor." Set against the backdrop of massive toy recalls that occurred throughout 2007, Ma's bill flew through the legislature in nine months, record time, and landed on the desk of Governor Arnold Schwarzenegger. For a while there was some nervousness as to whether he would sign it or not, but not even "The Governator" can stand in the way of irate moms with their toddlers and rubber ducks.

In October 2007 the continent's first bill restricting phthalates passed into law. "We must take this action to protect our children," said Schwarzenegger. "These chemicals threaten the health and safety of our children at critical stages in their development."

"This law is the product of the politics of fear," fumed Jack Gerard, then President of the American Chemistry Council. "It is not good science, and it is not good government. Thorough scientific reviews in this country and in Europe have found these toys safe for children to use."

But the chemical industry's phthalate headaches were about to get an awful lot worse. Within a few months of the California bill's passage, a dozen other states had introduced phthalate bans. Janet Nudelman laughed when she told me about ExxonMobil (the manufacturer of DINP, one of the most common phthalates in toys) hiring a "big contingent of lobbyists who just traveled from state to state. One thing that they did, which was particularly obnoxious," she said, "was to send lobbyists to Vermont to organize the sports fishermen against the proposed phthalates ban there. And you're probably thinking 'Why would sports fishermen be opposed to banning phthalates in children's toys?' Well what they were saying to these guys was that they would no longer be able to use their plastic wiggly worms, because the worms get their rubbery-ness from phthalates. They wouldn't be able to fish anymore."

Nudelman explained how, during the debate on the California legislation, the chemical industry used a Scrooge-type argument to defend their products. "They sent out nicely designed flyers to key districts throughout the State, the districts of moderate Democrats, which said: The legislature is trying to ban your child's beach ball." Just as in San Francisco, this hyperbole backfired and actually so annoyed some legislators that they ended up supporting Fiona Ma's bill.

During Washington State's phthalates debate, Nudelman told me, "the industry argued that Washington environmental health advocates were trying to cancel Christmas . . . Santa would be stopped in his tracks." Nudelman chuckled. "They were just using whatever they could come up with to create misinformation and confusion around what was really a simple piece of legislation banning six phthalates from some narrowly defined children's toys and childcare articles."

Despite this industry fearmongering, Nudelman remembered that in the autumn of 2007, there was "incredible momentum at the State level that literally catapulted the phthalates issue into the arms of Congress."

Easy Rider

There's a funny practice in the U.S. Congress that's rare or nonexistent in other countries. And that is the use of "riders"—additional legislative bits and pieces that members of Congress attach to bills already under consideration. Riders are usually created as a tactic to pass a controversial provision that would not pass as its own law. It's sort of the legislative equivalent of the remora: the tropical suckerfish that glues itself to sea turtles and sharks and is piggybacked to its destination.

It was a Californian, again, who started the ball rolling. Following hard on the heels of the phthalates ban passing in Sacramento, Dianne Feinstein, the senior Senator from California, introduced a federal Toxic Toys bill to Congress. She very quickly saw an opportunity to accelerate its progress according to Janet Nudelman. "Senator Feinstein saw a train that was moving through Congress very quickly, and that was the *Consumer Product Safety Commission Improvement Act.*" This was a major piece of legislation introduced in response to the huge flurry of toys from China that were being recalled. "Parents were going nuts about toy safety," said Nudelman. "So Congress introduced this massive bill to overhaul the CPSC and crack down on lead in products."

Feinstein didn't just see the train; she jumped on it. She attached her phthalates bill onto the CPSC Improvement Act as a rider. Surfing the same wave of public concern that had carried them to victory in California, the Breast Cancer Fund cobbled together a national coalition effort to secure the bill's passage, which included the National Council of Churches, the American Nurses Association, MomsRising.org and scores of environmental and public health organizations—about 60 groups in all. At one point Congress received a sign-on letter from 82 state legislators in 26 states, demanding action on phthalates.

Interesting fissures appeared in the industry opposition to the bill at the federal level. As opposed to California, where the toy industry and chemical industry were in lockstep, Hasbro, the second-largest toy company in the world, was actually supportive of a federal bill. The company clearly saw this as the lesser of two evils when compared to the unsettling prospect of different phthalate regulations in all 50 states. The American Chemistry Council and ExxonMobil remained the most vehement opponents of the federal ban.

After a remarkably short period of time, the CPSC reform package, complete with phthalates ban, passed on a voice vote in the Senate, meaning, for all intents and purposes, that it was unanimous. "All the more remarkable," said Nudelman, "if you consider that a few months earlier nobody had even heard of phthalates before." A Breast Cancer Fund press release at the time said, "This is a David and Goliath victory. Public health advocates and parents were up against big oil and the chemical industry, and we won."

It fell to George Bush to ink the deal. In August of 2008 he signed one of the most significant pieces of consumer protection legislation in a generation. It permanently prohibited the sale of children's toys or childcare articles that contain more than 0.1 per cent DEHP, DBP or BBP. The legislation took a precautionary approach to the other three phthalates—DINP, DIDP and DNOP—and imposed an interim ban while the science was studied further.

Importantly, the legislation put the ball in the industry's court to demonstrate that these phthalates are safe before they're allowed back on the market—the first time this has been done in U.S. law with respect to toxic pollutants. Veteran congressional observer Jane Houlihan, Vice President for Research of the Environmental Working Group, calls this "an important shot over the bow for the chemical industry." "The industry mobilized like crazy on phthalates and created working groups and armies of lobbyists to combat the proposed bans and restrictions," she said. "And in this case they lost the battle. It's a sign that people aren't willing to put up with these chemicals in products anymore."

Houlihan hopes that it's a sign of things to come: "In the U.S., the Toxic Substances Control Act requires that the Environmental

Protection Agency [EPA] prove that a chemical is harming health before they can take action. And what we need instead, and what the Feinstein bill establishes for phthalates, is a system where that burden of proof is reversed. Where companies who are making chemicals that are ending up in people's bodies, companies who are putting them in products, have to prove up front that their product is safe and their chemical is safe before it's sold. That's a pretty common sense idea, and it's time that we apply it to the other 82,000 chemicals currently in commerce."

Nearly a decade after toxic kids' toys were banned in Europe, this "better safe than sorry" approach finally came to America.

Better late than never.

The Ketchup Riddle

Of course, the phthalates story isn't over yet. This noxious class of chemicals is still in many household items, especially in the bathroom. Despite George Bush's amazing action to ban toxic toys in the U.S., some jurisdictions like Canada have yet to catch up. But thanks to the efforts of some great advocates like the Environmental Working Group and the Breast Cancer Fund, and a few determined legislators like Fiona Ma, there's some good news and some progress to celebrate.

A little late for my kids. But my sister's little boy, born shortly after George Bush signed the historic phthalates ban, won't have as many toxic toys to worry about.

The phthalates story illustrates many of the themes you'll hear again and again throughout this book. The extent to which the chemical industry will never admit there's a problem with their products. The determination of environmentalists and public health advocates to see that the right thing is done. The fact that making different choices as consumers can fundamentally affect the levels of toxic chemicals in our bodies but also that in the end, the only way to well and truly solve the problem of pollution in everyday items is for governments to get their acts in gear.

Here's a small anecdote to illustrate the sometimes Solomonic challenges faced by today's consumers. During the writing of this chapter, I took a late-night run to our local supermarket for a handful of items for the kids: milk, bananas, ketchup. My wife, Jennifer, is usually the one who does the groceries, so I typically take twice as long in the store as she does.

Make that ten times as long in this case as I stood blinking—staring—at the ketchup selection, honestly perplexed.

- The organic ketchup came in a plastic bottle.
- Aylmer ketchup, an old Canadian brand made from locally grown tomatoes, also came in plastic.
- The only option in a glass bottle was the non-organic, non-local Heinz ketchup.

What to do? The two choices that were either lower in pesticides or, by virtue of being local, easier on the carbon emissions, were undoubtedly higher in plasticizer chemicals. You can take it for granted that I definitely didn't buy one of the many non-organic, non-local, plastic-bottled brands. But beyond that I was suffering from ketchup-option paralysis.

That evening, anxious to get back home to relax with my sweetie, I chose to limit my family's pesticide intake. I headed to the checkout line with the organic ketchup in the certain knowledge that at the rate my kids ate the stuff—smearing it over their fingers, face and hair—I'd be able to make the same decision all over again in a few weeks' time.

Replicate my ketchup example many times over and you can see the scale of difficulty that the average shopper faces when trying to make choices to limit their family's exposure to toxic chemicals. Only government action can solve this, in this case with a regulatory decision to limit packaging with plasticizing chemicals or pesticide use or to buttress the production of local food.

As I drove home I reflected on the fact that only government leadership can provide us with the glass-bottled, locally grown, organically raised ketchup we all deserve and prevent any similar late-night supermarket conundrums in future.

THREE: THE WORLD'S SLIPPERIEST SUBSTANCE
[IN WHICH BRUCE TRAVELS TO TEFLON TOWN]

[Open on suburban kitchen, Wife and Husband arguing]

WIFE: *New Shimmer is a floor wax!*
HUSBAND: *No, new Shimmer is a dessert topping!*
WIFE: *It's a floor wax!*
HUSBAND: *It's a dessert topping!*
WIFE: *It's a floor wax, I'm telling you!*
HUSBAND: *It's a dessert topping, you cow!*
SPOKESMAN: *Hey, hey, hey, calm down, you two. New Shimmer is both a floor wax and a dessert topping! Here, I'll spray some on your mop . . . and some on your butterscotch pudding.*
> —"Shimmer," *Saturday Night Live*, 1975[1]

IT'S WELL KNOWN that Teflon—and its chemical relatives (called PFCs, or perfluorinated compounds)—are used to coat frying pans. Less well known is the fact that they're also used to line pizza boxes and windshield wipers and to make bullets and computer mice, and they're a key ingredient in cosmetics and clothing.

"It's Everywhere," says DuPont's tagline for Teflon. And that is precisely the problem. It's not supposed to be everywhere. Not in the flesh of ringed seals in the Arctic.[2] Not in the blood of 98 per cent of Americans[3] and certainly not—as we shall see—in the drinking water of the residents of Parkersburg, West Virginia.

PFCs are so prevalent and persistent and their uses are so varied, that they've made one of my favourite *Saturday Night Live* skits, which aired over 3o years ago, seem downright prescient. In the skit Dan Aykroyd is spraying creamy foam into his bowl while Gilda Radner is using the same product to clean the kitchen floor. The first time I saw it, I thought it was hilarious. Now it isn't quite so funny, because something very similar is happening with PFCs. They're synthetic chemicals that line our popcorn bags and are sprayed on our rugs and clothes to keep them stain free. They're used to put out fires, and of course, most famously, they line frying pans to prevent our food from sticking. DuPont is no longer the only one that knows Teflon is everywhere; now the U.S. Environmental Protection Agency (EPA) knows too. The persistence and global extent of the Teflon manufacturing ingredient called PFOA (short for perfluorooctanoic acid) has made the EPA sit up and take notice. And PFOA is now at the centre of a $3oo million legal battle.

The Chemical That Lasts (Almost) Forever

PFOA is one of many PFCs manufactured by DuPont and a hand-ful of other companies. The well-known non-stick coatings Teflon, Silverstone and Capstone are all DuPont brand names for various types of perfluorinated compounds. Teflon has the honour of being listed in the *Guinness Book of World Records* as the world's slipperiest substance. In addition to their prevalence in our kitchens, PFCs are also common in our bedroom closets: Gore-Tex is a brand name for PFC water-repellent fabric, as are Scotchgard and STAINMASTER.

Until recently Scotchgard contained a chemical known as per-fluorooctane sulphonate (PFOS). But in 2000, 3M, the maker of Scotchgard, voluntarily removed this ingredient from its products when it discovered how persistent it was in the environment. In a notice to Environment Canada, the U.S. Environmental Protection Agency advised Canadians of the voluntary phase-out, stating that PFOS "appears to combine persistence, bioaccumulation and toxicity properties to an extraordinary degree."[4] PFOS remains widespread in the environment.[5]

Like its chemical cousin PFOS, perfluorooctanoic acid (PFOA)— or C8 as it's also known because of the eight carbon atoms in its molecular structure—has lots of problems. It is considered by many scientists to be toxic and to cause birth defects, developmental problems, hormone disruption and high cholesterol.[6] The EPA has labelled it a "likely carcinogen,"[7] and it's now found in every corner of the globe.[8] Yes, it has many unique and useful chemical properties. It is amazingly durable, it resists other chemicals, it is fireproof and nothing sticks to it. But the very properties that make PFOA commercially desirable also cause environmental and human health concerns. Its durability, slipperiness and resistance to breakdown are a major problem. Nothing gets rid of it. Not sunlight. Not our stomach acids. Once PFOA is created it takes a very, very long time to go away. It may persist in the environment for centuries. Every molecule that has ever been created is still around and will be around for the foreseeable future. It's as though our bodies and the environment were a giant United Way thermometer, with a red line representing PFOA slowly and steadily inching its way up until we're full of the stuff.

DuPont is the sole U.S. manufacturer of PFOA, which is used to make Teflon and many other products in a town in West Virginia called Parkersburg. Over the years quantities of the chemical have escaped into the local air and water, and PFOA-linked health concerns have become a major local issue.

In addition to being directly emitted into the environment, PFOA is also indirectly created as a breakdown product of a specific kind of PFCs called "fluorotelomers," which are sprayed onto fabric or furniture like sofas by consumers, or are factory-applied. Danish researchers recently found that levels of PFOA and other perfluorinated compounds in polar bears have increased by 20 per cent or more since 2000.[9] Scott Mabury, a chemist at the University of Toronto and one of the world's leading researchers in this area, has found the chemical in many species and ecosystems.[10] I spoke with some of his research team, and their conclusion is that this ongoing and increasing pollution can be explained only by the fact that the fluorotelomers found in various products degrade into PFOA.

Mabury described the process in a 2006 interview with the Canadian Broadcasting Corporation:

> What we've discovered is that we can measure in the atmosphere fluorotelomers (also known as fluorinated alcohols) that we think are escaping from a variety of consumer products: carpet, coatings for water and stain repellency. Those fluorinated polymers that are used to coat that carpet, some of the residual material is escaping into the atmosphere. We can measure it all across North America. . . . We know it will last in the atmosphere about 20 days, sufficient time to be transported to really remote areas, like the Arctic. We know that Mother Nature, in trying to rid itself of chemical pollutants, transforms that fluorinated alcohol into much more persistent and bio-accumulative materials called perfluorinated carboxylic acids, things like PFOA. . . .
>
> There's not much PFOA itself in these consumer products, but the precursors are there. And I guess it was one of our contributions to actually make that linkage. It's not PFOA itself moving around the world; it's the pre-cursor. The travel agent is the fluorinated alcohol—that's the one that has sufficient volatility, vapour pressure. It escapes in the atmosphere, moves long distances in the atmosphere, before being transformed into PFOA, and then potentially moving up the food chain.[11]

Manufacturers dispute this theory. They believe the chemicals in the coatings are stable and not getting into the environment the way Mabury describes.

Parkersburg, then, is at the centre of the PFOA story. We can all likely think of more than enough examples of people being polluted by the chemical factory or toxic waste dump "next door" (think Love Canal or John Travolta in *A Civil Action*). And this is one of the dimensions of the Parkersburg experience. But the tale of Parkersburg may be the first environmental-disaster story in which a small town is also responsible for contaminating the entire world and almost every living thing in it.

I had to see it for myself.

The Tennants of Teflon Town

Parkersburg, West Virginia, is the town that Teflon built. It's here that DuPont has long manufactured this well-known product.

I arrived at Wood County Airport on an unusually hot September morning, hopped into my little red rental car and headed into town. The temperature on a sign at one of the local banks read an even hundred degrees Fahrenheit. It was so hot I couldn't help thinking of the proverbial egg frying on the sidewalk. And of course, in Parkersburg, you wouldn't have to worry about it sticking.

The town has a picturesque setting at the confluence of the Little Kanawha and Ohio rivers. The drive into town took me along winding roads through lush Appalachian forests. The area seemed sparsely populated, with quaint farms located in the valleys. The combination of forests, arable land, coal and the rivers made it a natural location to support early industrial development. First settled by Europeans in 1772, Parkersburg is one of the oldest towns in North America and the oldest in the Mid-Ohio Valley. In 1860 it was the site of the first producing oil well in the state, and by the late 19th century, Parkersburg was a bustling merchant town with tanneries and ship-building facilities. As with many rapidly growing industrial towns of the era, a small number of people controlled everything: the land, the courts, the industry. Judge John J. Jackson, for example, the longest-serving federal judge in U.S. history, was involved in almost every major industry in Parkersburg in the late

19th century and brought to West Virginia a strong pro-business perspective that continues today. For those unfamiliar with the term, a "pro-business environment" is political code for a place where environmental and labour protection is not a priority. West Virginia is well known for upholding this reputation.

Parkersburg is both ground zero for the global fallout of PFOA and the centre of an Erin Brockovich–like legal battle. A mile or so downriver from downtown Parkersburg is DuPont's Washington Works chemical plant. It is located on what are called the "bottom-lands"—the floodplain adjacent to the Ohio River. The bottom-lands were considered to be the most desirable of all land holdings in the late 18th century, when veterans of the American Revolution were given the land promised to them for serving their new country. Perhaps the best-known American of all time, George Washington, was given several thousand acres of prime land on the Ohio River, and he planned to make it his private retreat. Though this never materialized, his name has lived on in the Washington Works plant, which is no bucolic estate. It's two thousand acres of sprawling roads, railways, pipes, tanks and chemical chimney stacks. The Works' neighbours include facto-ries, oil refineries and coal-fired power plants.

Directly across the river from the DuPont chemical factory is the drinking-water well field for the town of Little Hocking, Ohio. As I stood in the well field, the DuPont facility dominated the view across the river. I could see pipes going in every direction, includ-ing under the river, and emerging on the Ohio side, where I was standing. The newly constructed water filtration system was in place and easily visible. It had been installed in 2006 to remove contaminants from the Little Hocking water supply as part of a legal settlement with DuPont.

The Tennant farm, located a short distance from Washington Works, was a typical family farm—until the cattle began to waste away and die. Jim and Della Tennant farmed for many years on their beautiful acreage, raising their extended families and their cattle until things started to go awry in the early 1980s, shortly after they sold a portion of their farmland to DuPont.[12] According to Callie Lyons, a local journalist and the author of *Stain-Resistant, Nonstick, Waterproof and Lethal,* a book that chronicles the story of the C8 contamination in the Parkersburg area, the Tennants started to notice strange things on their property: Local wildlife were dying and minnows had disappeared from their stream, the same stream their cattle drank from. They watched helplessly as their animals died, unable to assist because they had no idea what was causing the problems. In addition to those wasting away, a number of cattle were born with serious abnormalities.

By the late 1990s the Tennants' herd was decimated, and worse, family members were now suffering from respiratory ailments and various cancers. In an interview for National Public Radio, Della Tennant described the scene of a dying cow: "It had the most terrifying bawl, and every time it would open its mouth and bawl, blood would gush from its mouth. And there was nothing you could do. It was suffering, and there was nothing you could do. And whenever you think about feeding all those animals to your children, all the time they were growing up, it's something that puts a lump in your throat you can't take away."[13]

The Tennants discovered that the piece of land they'd sold to DuPont was being used as a dump for hazardous waste from the Washington Works plant, and they would later find out that it was heavily contaminated with PFOA. Working with Rob Bilott, a Cincinnati lawyer with roots in Parkersburg, they sued DuPont for damages. DuPont settled privately with the Tennants in 2001. The details of the settlement are not known. It certainly helped keep PFOA contamination—temporarily—out of the public eye, but it didn't keep it out of the water or out of the bodies of citizens.

Joe

The real story of Parkersburg is not about a single contaminated farm; it's about widespread chemical contamination in the local watershed. It all started when Joe Kiger, along with the other residents, received a letter from the Lubeck Public Service District (LPSD), advising them that a substance called PFOA had been found in their drinking water.

I met with Joe in the oak-panelled library of the historic Blennerhassett Hotel in downtown Parkersburg. He had just come from coaching the high school football team, and it was obvious that his laboured breathing was a symptom of asthma. He confirmed that he did indeed suffer from asthma and went on to tell me he has had eight prostate biopsies and has five stents in his heart and a liver condition. Joe is convinced his poor health has something to do with the chemical contamination in Parkersburg. "How else can you explain all this disease in me and my friends?" he asked.

Joe was a big guy but fit. He looked very much the part of a high school football coach, sporting a full head of grey hair cropped short with a flat-topped brush cut. He looked as though he could have walked out of the pages of a 1950s high school yearbook. Joe has never worked for DuPont, but he was born and raised in Parkersburg and has had a long and diverse career, doing everything from construction to teaching to administering health and safety standards. He has more than a passing interest in PFOA. He is the lead plaintiff in a class action lawsuit against DuPont. Filing the lawsuit was something "I just had to do," Joe told me. "What good are you if you can't stand up for yourself?" he said. You could tell from his passion and commitment that he takes his role as a citizen very seriously.

Joe sat back and proceeded to tell me the story of how he first got involved in the largest environmental class action suit in U.S. history. The letter from the Lubeck water utility contained an assurance that according to DuPont, PFOA levels in the drinking water were safe. This was a red flag for Joe. Why, he wondered, was DuPont telling the water utility that the water was safe? Shouldn't it be the other way around? Joe didn't act immediately, but he held onto the letter. Over the next several months, a combination of things piqued his interest. Many of his friends were also suffering from health problems. Stories about the safety of chemical use at the DuPont plant were popping up in the news, and word of the Tennant farm lawsuit was spreading around town. It dawned on Joe that something was up: "I said to the wife, 'Where's that letter?'"

Joe Kiger is not the kind of guy who simply accepts what he hears, and he could tell that something wasn't quite right. He began calling around, starting with the Lubeck water utility, which he said was evasive at best. Then he got in touch with the West Virginia Department of Natural Resources and the Department of Environment. According to Kiger they all treated the issue "like the plague." Nobody would touch it. DuPont, he said, was of little help. The more he called, the more frustrated he became. According to Kiger, "it was like a wall went up." As he talked I was reminded of the famous Confederate General "Stonewall" Jackson, who was from around these parts.

After getting no help locally, Joe called the U.S. EPA regional office in Philadelphia. They asked him to fax a copy of the letter he'd received from the Lubeck Public Service District. The letter referenced the fact that the U.S. EPA regulates "the amount of certain contaminants in water provided by public water systems." The Lubeck Public Service District went on to say that "PFOA is unregulated for drinking water purposes" (that is, it has no limit established by a regulatory agency). Instead, it noted that "the DuPont Company has established its own drinking water guide-line" and "DuPont has advised the District that it is confident these levels [of PFOA found in Lubeck drinking water] are safe." Joe was perplexed. *How is it that DuPont can set a drinking water standard itself and then tell the water utility that the water is safe?* he wondered. Joe contacted the EPA because it just didn't sound right. An EPA employee said to him, "What the hell is PFOA doing in your water?" This employee sent Joe some information that included the name of a lawyer: Rob Bilott, the one who'd taken on the Tennant farm case. After a lengthy conversation with Kiger, Bilott agreed to take on the case as a class action suit against DuPont, with Kiger as the lead plaintiff.

The class action was filed August 30, 2001. The lawsuit alleged, among other things, that the actions of DuPont, together with those of the Lubeck Public Service District, "were conducted with such intentional, malicious, wanton, willful, and reckless indifference to the Plaintiffs . . . and flagrant disregard for the safety and property of Plaintiffs . . ."[14] that they were liable for punitive damages.

The lawyers didn't mince their words, although the plaintiffs in the class action suit were well aware of the important economic role DuPont played in the community. They made it clear from the outset that their goal was not to shut down the DuPont plant but merely to get answers to their questions and fair compensation. The plaintiffs wanted to know whether the PFOA in their water was a health problem or not.

As Joe said, "It was like eatin' possum: The more you chew, the bigger it gets." His decision to act as the lead plaintiff in the civil action was not taken lightly, and the ramifications in his life are still being felt today. Joe endured various forms of abuse from members of his community, which was quietly divided between those who were loyal to DuPont because they worked there or had family members working at the plant (known locally as DuPonters) and those who wanted to know more about perfluorinated compounds (PFCs) and their health. Some of the DuPonters were vocal and intimidating. Objects were thrown at Joe's house and harassment was always in the back of his mind. When one of Joe's neighbour's house accidentally caught fire and burned to the ground, a friend of his quipped, "They must have gotten the wrong address."

DuPont provides nearly two thousand direct, high-paying jobs in the community and at least that many indirect jobs, so the idea of a group of citizens threatening the future of the plant was considered to be tantamount to treason. I spoke with Lisa Collins, a local public relations consultant who worked on the court-approved health-data-gathering project in Parkersburg that included local blood testing. According to Lisa, "West Virginians like to keep to themselves," so the idea of a major class action suit against the town's largest employer was not a comfortable situation for many. People suspected that the plant was "killing us and feeding us at the same time," but nobody could imagine DuPont gone. "Nobody wants that," she said. Callie Lyons, the local journalist who wrote *Stain-Resistant, Nonstick, Waterproof and Lethal,* attributed this mindset to West Virginia's coal-mining history, which led people to accept "that having a high-paying job often meant getting sick." Such is the conundrum faced by so many one-industry towns, especially in areas where high-paying jobs are otherwise scarce.

Health issues in Parkersburg and at the Washington Works plant were not well known prior to the class action suit. They were a well-kept secret by DuPont. A little too well in the opinion of the U.S. EPA, leading the agency to sue DuPont for failing to disclose the fact that it had found PFOA in the blood of its workers at the Washington Works plant as early as 1981. The lawsuit was settled in December 2005, and DuPont was fined $16.5 million, the largest administrative penalty the EPA has ever obtained under U.S. law.[15] On seven of the eight counts, DuPont was found to have violated "the requirement that companies report to EPA substantial risk information about chemicals they manufacture, process or distribute in commerce."[16] [17]

A Discovery at the Plant

It turns out that DuPont knew of health risks associated with PFOA as far back as 1961, when company researchers discovered that rat livers were enlarged when exposed to very low doses of PFOA. An internal DuPont memo on PFOA and related chemicals advised that "all of these materials . . . be handled with extreme care. Contact with the skin should be strictly avoided."[18] DuPont maintains that levels of PFOA are too low to raise health concerns. 3M is another major corporation that manufactured and used PFOA and conducted health studies on its workers (but has since phased out the production of PFOA, PFOS and PFOS-related products). One study found that 3M workers directly exposed to PFOA were three times more likely to die of prostate cancer than the least exposed workers.[19] 3M discounted the results because the death rates were still within the average range for unexposed men, and in fact, they used the study to claim that PFOA was safe despite the cancer deaths.[20] The health effects are complex, the doses are very low and the human studies involve small numbers of people, so no single study on its own provides conclusive evidence of harm to humans. But putting all the information together, the weight of the evidence suggests that PFOA is a big problem. It is this assessment that led the EPA to deem PFOA a "likely carcinogen." Yet DuPont remains confident that PFOA is safe, stating: "DuPont believes the weight of evidence indicates that PFOA exposure does not pose a health risk to the general public."[21]

I headed off to Charleston, West Virginia, to meet with Harry Deitzler, the lead litigator in the class action suit against DuPont. Like Joe Kiger he was born and raised in Parkersburg. Harry is a former District Prosecutor who clearly has a passion for his work and a deep respect for the hard-working people of Parkersburg. He described the situation of the local citizens with a conviction and eloquence that helped explain his success as the country's leading PFOA prosecutor.

Harry told me about the DuPont study in 1981 in which the company was looking to see if there were birth defects among the children of any workers who'd been exposed to PFOA. Studies of rats showed very specific facial birth defects: cleft palates, nostril deformities and, according to Harry, an unmistakable tear duct deformity that seemed unique to PFOA.[22] DuPont chose eight women who worked with PFOA. If one birth defect was found among the eight women, Harry claimed, DuPont was planning to chalk it up as "coincidence," even though one birth defect in a sample size of eight would be highly unusual. If two or more birth defects were found, DuPont would know they had a problem on their hands. It turned out that two of the eight women did have children born with birth defects. Not only were the two birth defects similar, but they were the identical kind of eye and facial problems found in the studies of rats exposed to PFOA. DuPont's response? Transfer the women to another part of the factory, cancel the ongoing health studies and keep the results a secret. Harry's distaste for this episode was visible. Perhaps these were the acts of "reckless indifference" and "flagrant disregard" he had in mind when he used that language in the class action lawsuit.

For Joe Kiger, the lawsuit was about basic human decency. "They knew [PFOA] was in the water, they knew it caused deformities, they knew of the problem and they knew how to solve it." He wonders how, given the high rates of autism and asthma, the cancer clusters and the birth defects reported in the Parkersburg area, DuPont could keep doing what it was doing and not tell anybody.

Settlement

In early 2003 the Lubeck Public Service District settled privately, leaving DuPont as the sole defendant in the lawsuit. Following the judge's order to seek mediation, DuPont and the legal team of Deitzler and Bilott began working toward a settlement. By November 2004 they had reached an agreement in principle to settle out of court, and the settlement was finalized in February 2005. Over the course of the three-year civil action, Harry Deitzler estimated that they reviewed more than 1.5 million pages of documents, and the legal team racked up fees and related costs of over $22 million, all of which were paid by DuPont.

The legal settlement included a $71 million health and education project, the installation of a $15 million state-of-the-art water treatment facility to allow the six local water districts to remove PFOA from the water supply to the "lowest practicable levels" and the creation of a $20 million Science Panel to determine whether there is a probable link between PFOA exposure and adverse medical effects.[23] If the Science Panel does find this link, DuPont must pay up to $235 million to cover medical costs for any plaintiff who can demonstrate direct personal harm.[24] The establishment of the Science Panel is a novel approach to the settlement, and it helped Harry Deitzler and his firm win the coveted "Trial Lawyer of the Year" award in 2005 from the Trial Lawyers for Public Justice Foundation.

Adding these costs up DuPont agreed to a possible $340 million settlement, one of the largest environmental class action settlements known. To put this amount in context, however, it is roughly equivalent to only one-tenth of one year's after-tax profit for DuPont, based on over $3 billion in profits and revenue of $30 billion in 2007 alone.[25]

For Joe Kiger and others who wanted to know whether the PFOA in their water was harming them and their families and friends, the Science Panel has become a central part of the settlement. It is a "bullet proof" science panel, consisting of some of the world's leading epidemiologists and set up in such a way that DuPont, Kiger and the other plaintiffs all had the ability to veto any expert that they think might have a bias. As a student of science policy, I find this a brilliant approach, since the determination of probable harm is to be made by experts, not by a jury of peers. The process could cut either way. Scientists are often extremely cautious, and this could play to DuPont's advantage. But given that the panel is tasked with finding a "probable link" between PFOA and health problems, there is no need for them to make absolute statements. Whatever they determine will no doubt be taken as the *de facto* truth. So the findings will have important global implications.

The panel, which was created in February 2005, has a clear mandate with two very specific objectives. The first is to determine whether there is "an association" between PFOA and disease generally. The second is to find out whether there is a "probable link" between PFOA in the water and disease in Parkersburg.

According to Harry Deitzler, if the Science Panel makes a positive finding related to a "probable link," in legal terms DuPont will not be able to deny "general causation." This means that DuPont will not be able to deny that PFOA probably causes harm. They can still deny "specific causation," however, meaning that a specific individual claim can be denied by DuPont if they can prove that particular damage is not related to PFOA. Each individual case will therefore need to be considered separately.

If a "probable link" is not established by the Science Panel, then those claiming to have health problems as a result of exposure to PFOA at the Washington Works plant will not be successful in bringing their claims against DuPont. Needless to say the future of many individuals from Parkersburg will be greatly affected by the findings of the Science Panel. Because the Environmental Protection Agency classifies PFOA as a "likely carcinogen" and numerous animal studies have demonstrated direct links between PFOA and disease, the first task of the Science Panel has largely been accomplished already.[26] Harry Deitzler is confident that the panel will come to this conclusion. But it's much less certain that the panel will discover a "probable link" between the PFOA in the local drinking water and disease or birth defects in the local population.

Harry was heavily involved in the details of the settlement, including the selection of panel members and the development of the data-gathering strategy. In a traditional study a sample of the population would be used and results would be extrapolated to the entire population. But Harry suggested that every citizen in the Parkersburg area should participate by donating blood. This approach was unprecedented in epidemiology. The experts were skeptical that one hundred per cent participation could be achieved, but Harry had been involved in similar situations and he knew that public participation could be motivated by cash. The question, Harry said, was simply *How much would it take to get every citizen to donate their blood?* "I figured I'd do it for four hundred dollars," he told me. "And if I would do it for four hundred dollars, then surely anybody would," he said, alluding to his above-average wealth as a trial lawyer. In the end local citizens were offered four hundred dollars to donate their blood, and a total of 69,800 samples were taken. Harry figures that covers just about everyone in town.

The Parkersburg case is now considered to be the most comprehensive community toxicological epidemiology study ever conducted. The three members of the Science Panel agreed to by Harry and DuPont's lawyers are internationally recognized epidemiology experts. Although the Science Panel concept is a cutting-edge development in the legal and health science professions, it comes with its own drawbacks. Some local residents view it with suspicion, and the timeline is frustratingly slow, with at least four years of study involved. The panel's task is also a daunting one: It is overseeing 11 separate health studies covering everything from immune, liver and hormone disorders to geographic patterns of cancer to birth outcomes.

Though the final study will not be completed until 2011, the first indications of where things may be headed came to light in the panel's preliminary results, released in October 2008. The panel found that study participants with higher levels of PFOA in their blood had higher cholesterol levels. Participants also had nearly six times as much PFOA in their blood as the average American. "We can't conclude anything yet about a possible link between C8 [PFOA] and disease because we don't know whether the disease came first, or C8 came first. Our charge is to determine if there is a probable link between C8 and any disease," said Science Panel spokesman Kyle Steenland.[27]

Before leaving Parkersburg I stopped at the Crystal Café for breakfast. I was sitting alone at the chrome-and-Formica-clad counter when I realized my young waitress was singing the title words from the early 1980s Thomas Dolby song "She Blinded Me with Science" (which, as you may recall, also featured the repeated lyrics "Science!" and "I can smell the chemicals"). Given that the findings of the Science Panel will have such a dramatic effect on the lives of the people of Parkersburg, I can't imagine a more appropriate soundtrack for my West Virginia visit.

In fact, I'm not sure there's ever been another community in history whose future is so completely and directly bound up in "science" and the dramatic struggle over how best to interpret its findings.

Out of the Frying Pan . . .

The non-stick frying pan is surely the icon of the world of perfluorinated chemicals, like those produced at the Washington Works plant. In centuries past canaries were lowered into coal mines, and if they died, miners knew that the air below ground could be toxic to them as well. Perhaps we should heed the modern-day equivalent: Non-stick coatings literally kill canaries in kitchens.

It seems that the delicate respiratory systems of birds cannot tolerate the fumes from non-stick pans when they are heated to high temperatures. Their little aviator lungs hemorrhage, becoming filled with fluid and causing them to drown.[28] This rapid and deadly syndrome has been known for 35 years and even has a name: Teflon toxicosis. Non-stick frying pans, toaster ovens, cookie sheets and pizza pans have all been implicated in pet bird deaths. And the bird killings are not restricted to cooking devices. Irons, space heaters, carpet glues and new sofas have also destroyed the sensitive lungs of pet birds, causing them to suffocate. More than one incident of mass bird deaths has been reported in the vicinity of non-stick-coating manufacturing plants in Canada and Great Britain. And there are also reports of birds dying from self-cleaning ovens, heat lamps and oven interiors with non-stick coatings.

As you might imagine this is a touchy subject for the manufacturers of these chemicals. Nobody likes to be called a pet killer. And having one of your prized brands followed by the word "toxicosis" is not exactly a marketing dream come true. Non-stick companies such as DuPont are quick to point out that under "normal" cooking conditions, the pans should not kill your cockatoo. DuPont gives this advice: "Sadly, bird fatalities can result when both birds and cooking pots or pans are left unattended in the kitchen—even for just a few minutes. Cooking fumes from any type of unattended or overheated cookware, not just non-stick, can damage a bird's lungs with alarming speed. This is why you should always move your birds out of the kitchen before cooking."[29] The advice from DuPont does not mention Teflon.

Of course, if you move a small bird out of the area where fumes might harm it, it will be less likely to die. However, the question remains: Do non-stick cooking utensils pose a greater and substantial risk of possibly causing health problems for the humans who use those utensils? The jury is still out on this, but there's some convincing evidence to suggest that non-stick coatings can heat up quickly to levels where numerous toxic gases (some carcinogenic to humans) can be released. According to manufacturers of non-stick coatings and independent studies, the coatings start to break down somewhere between 400°F and 550°F, and highly toxic fumes are emitted when the pans are heated above 680°F.[30] Further, they claim that these temperatures are rarely reached under normal cooking conditions. This is where the disagreement, and contrary evidence, lies. Independent studies commissioned by the Environmental Working Group (EWG) show that pans with non-stick coatings can reach well over 700°F in five minutes or less when preheated on "high" on a conventional electric stove.[31] If this represents "normal cooking conditions," toxic fumes may always be released and non-stick coatings will always break down under normal cooking conditions.

According to DuPont's own studies, toxic particles start to form above 464°F, and at 680°F Teflon-coated pots and pans release numerous toxic gases, including known human carcinogens.[32] There are now an increasing number of non-stick products (including broiling pans and oven interiors) that are used in such a way that they guarantee that higher than "normal" cooking conditions will be reached. This is because broiling is a much higher-temperature method of cooking and when an oven interior is heated, the surface temperature of the oven becomes much hotter than a pan with food in it. The EWG report notes that non-stick stovetop drip pans (the trays that fit under the elements on electric stoves) can reach temperatures as high as 1,000°F.[33] At temperatures this high the non-stick coatings break down further and the noxious gases that are released include perfluoroisobutylene (PFIB), a relative of the World War II nerve gas phosgene. That ought to do Polly in if the toxins released at "lower" temperatures don't.

Although birds appear to be the most sensitive species, they are not the only ones affected by heated non-stick coatings. In more than one experiment, non-stick pans heated to 800°F killed a group of rats within four to eight hours.[34] In several cases of bird deaths after exposure to fumes from Teflon, the bird owners were also hospitalized with what is known as "polymer fume fever," which causes flu-like symptoms, including difficulty breathing, accelerated heart rate, chills and body aches.[35]

In another strange twist, smoking in the presence of Teflon is exceptionally toxic. Minuscule Teflon particles can decompose in a burning cigarette, causing polymer fume fever in smokers who work in non-stick coating factories. It's not known to what extent problems are caused by smoking at home while heating non-stick pans, but since both practices are fairly risky on their own, it might be wise to avoid that butt while sautéeing your mushrooms in a non-stick pan.

PFCs are part of the modern-living movement, where cost trumps quality and convenience trumps all. It may be a little harsh to say, but these products seem to be marketed with the assumption that consumers have no time or interest in learning to cook and eat well. On the surface, non-stick coatings in pots and pans sound fine. The classic image is a fried egg sliding effortlessly out of a pan onto a plate. No spatula needed! The pan is then wiped clean with a cloth and is ready to use again. Sounds perfect. Simple, easy. Even the most culinarily challenged can now fry an egg. But no serious chefs I know would choose a non-stick frying pan, and this has more to do with the heating properties and browning quality of frying pans made of stainless steel, copper or cast iron. If you know what you're doing, it's not difficult to get your egg to slide out of a heavy-gauge stainless-steel or copper pan with similar ease.

I happen to think I'm pretty handy with a frying pan, so here are a few very specific pieces of advice from my own cooking experience on how to kick non-stick.

First, you do need to buy a decent frying pan. Your pan doesn't need to be a high-end gourmet item, but it must have a reasonably solid base, so it can heat quickly and evenly and retain heat at a constant temperature. Even if you spend a bit more per pan, it will actually save you more money than if you buy less expensive, non-stick pans every few years after the coating has been scraped off and consumed with your scrambled eggs.

There are three basic categories of pots and pans to consider (and many variations on these): cast iron, stainless steel and enamel-coated cast iron. My favourite, and the all-American classic, is the basic black cast iron skillet. The best ones in my experience are made in the U.S. The beauty of cast iron is that if it is treated properly and cared for, it outperforms non-stick. The only things that you should not cook in cast iron are high-acid vegetables such as tomatoes, because they concentrate too much iron if they are cooking for a long time.

There are three main reasons food sticks to a pan. First is that the pan is not hot enough. (And a pan must reach the correct temperature *before* any food is placed in it.) Second is not having a nice coating of oil in the pan. And the third is using a plastic spatula instead of a metal one. (Plastic spatulas tend to act more like shovels than spatulas.)

So follow this simple advice: make sure the pan is hot enough and that it has a nice coating of oil—and use a metal spatula—*et violà!* You can relegate your non-stick pans to the dusty back of your cupboard.

A Sticky, Stain-Ridden Future

Working as an environmentalist for the past 20 years, it has often been frustrating to look back and see so much intelligent analysis, developed by environmentalists for the benefit of society, be largely ignored by government and industry. In the early 1970s environmental health experts developed, for example, the concept of assessing chemicals as persistent, bioaccumulative and toxic (PBT) as a gauge for targeting substances to be phased out. The rationale for using PBT analysis is fundamentally sound from scientific, ecological and ultimately economic perspectives. When substances that are found to exhibit all three of these properties are in widespread use for decades, one can almost be certain that human health and/or ecological problems will surface, and in addition to the human health costs, they often result in costly legal settlements.

The whole point of developing such solutions for PBTs is to avoid potential ecological, health and economic damage, including possible catastrophes. Despite this knowledge and the efforts to apply it to PFCs, pesticides, mercury and a host of other chemicals, the typical approach of government and industry is to do too little, too late. Corporate interests often set the stage for legislative and regulatory frameworks in North America and hamper the efforts of well-meaning regulators.

I was sitting in my little red rental car before departing from the historic but slightly dog-eared town of Harmar, Ohio, just up the river from Parkersburg. It was late in the day and several merchants were chatting on the sidewalk after closing up their antique shops. I was not paying much attention to their conversation, but oddly enough I overheard one of them say, "You're soaking in it," and then another made reference to Madge's whereabouts. Madge, of course, was the fictional manicurist from the Palmolive commercials who, for a 25-year period, would tell her customers not to worry about the fact that they were soaking their hands in dish detergent. Her famous line "You're soaking in it!" is apropos for Parkersburg residents and PFOA. It also correctly describes the prevalence of many other toxic chemicals in all of our lives.

Dish detergent is a mild (in fact, "more than just mild") proxy for many of the chemical exposures we face today. We are eating, drinking and soaking in tens of thousands of potentially toxic substances, most of which we know little about. When all of these chemicals are combined together in our drinking water, we know virtually nothing about how they interact with each other or our bodies, and we know particularly little about how they affect developing brains and fetuses. Medical experts have no idea whether 20 different carcinogenic chemicals in the water are 20 times more likely to cause cancer or 100 times or whether their effect is likely to be no different than that of just one chemical. In fact, this area of study, called "cumulative risk assessment," is still in its infancy, and studies of the health effects of multiple chemical exposures are rarely undertaken due to the complexity and lack of scientific understanding of chemical interactions in humans. It is hard enough to understand the effects of one chemical, because it often takes 50 to 100 years of use and study to figure that out.

Today, we are told by real people who work for chemical companies (not fictional manicurists) not to worry, even though we are essentially soaking ourselves, our furniture, our clothes, our food, our packaging and most other things in our lives in a toxic brew, with unknown consequences. We are also soaking our bodies in it in our bathtubs and drinking this mystery concoction in our tap water.

Tap water, sadly, is where the legacy of toxic chemical manufacturing emerges in its most nefarious form. PFOA is no different. The storylines of small American towns contaminated by industrial pollution are remarkably similar in the best-known cases. They start when citizens notice unusual health effects in the community and raise concerns with local officials. It is an all-too-familiar tale, and I often wonder how many of these stories are happening every day around the world where either citizens have no legal recourse or where there simply isn't a strong enough leader to fight the battle. What if there is no Erin Brockovich or Lois Gibbs or, as in the case of Parkersburg, Joe Kiger? Behind the community leaders there are organizations such as the Environmental Working Group in the United States and Environmental Defence in Canada, spearheading campaigns against toxins. Without their work it is unlikely the public would have any idea of the potential health risks these chemicals pose.

The days are numbered for PFOA and many of its PFC relatives, thanks to increased consumer, environmental and community vigilance. Despite DuPont's claims that PFOA is safe, the company is phasing out its manufacturing, use and purchasing of the chemical by 2015. Does this mean that society is no longer willing to accept the modern-living dreamworld, where spills bounce off of fabric and fried eggs glide onto plates as though they were self-propelled, if it means compromising their children's health and polluting the planet? That remains to be seen. You can rest assured that PFOA is not the last of the PFCs we will see. Industrial uses are exempt from the voluntary phase-out, and DuPont is already marketing a PFC-based non-stick alternative called Capstone, which according to DuPont, offers "the same or better performance you have come to expect, without a compromise in fluorine efficiency."[36]

Off-Gassing and Us

Whenever we tell people about this book, we're often asked whether our self-experimentation made us feel ill in any way. Did our bodies respond in a noticeable fashion to the chemicals we were subjecting them to?

Mostly, the answer was no. We couldn't pinpoint any symptoms related to mercury, bisphenol A or triclosan. Rick felt a bit overwhelmed by the smells of the many overlapping fragrances of the personal-care products and air fresheners he used. But this was comparable to the surprise that most people feel when they get too close to the odiferous perfume counter in a department store. The only experiment that hit us both really hard was the one involving perfluorinated chemicals.

Ironically, it was also the one that—in the unique way we were defining success—flopped.

The experiment was simple. We wanted to see whether we could increase the level of PFOA in our blood. After speaking to a few experts, we decided to try measuring the effects of inhaling the off-gassing from a "normal" stain-repellent treatment. We needed a dedicated space, and I had an unused room in my condo that fit the bill.

The test room was 3 metres by 3.7 metres, with a large walk-in closet and a small hallway. It was 2.4 metres in height. We furnished it with the local thrift shop's finest: a wall-to-wall beige carpet remnant, a pink loveseat, a brocaded easy chair and gold curtains. Man, it was ugly, but the furniture suited the 1970s vintage condo and added to the feeling that we really were living in the heyday of modern chemistry. The loveseat looked as though it could have used a good thick coating of stain repellent 20 or so years before.

Once we had the room ready, we booked an appointment with a carpet professional we found through the Yellow Pages (there were lots and lots of companies to choose from). The guy arrived at the condo with a canister on his back, full of what he claimed was Teflon Advanced. With no breathing equipment of any kind (I hate to think what his PFOA levels are), he sprayed the carpet, chairs and curtains just as he would for any client. As we learned in our chats with a variety of carpet treatment companies, many people have their furniture and drapes spritzed with stain repellent, pre-sumably to ward off those flying glasses of red wine from overly enthusiastic New Year's Eve toasts and other such antics. The whole procedure cost sixty dollars.

We asked the guy as he left how long we should air the room out before starting to use it again. "About 20 minutes should do it," he replied.

Not a chance. The stain-repellent stench after 20 minutes was completely over the top and far too much for Rick and me to handle. The intense chemical odour—not unlike a really reeking dry cleaning shop—caught in our throats and made our eyes water like crazy. It was more than two hours before we could spend any time in the room, and even then Rick and I took turns in the early part of the experiment poking our heads out the door to gulp some fresh air.

For two days we sat in the room with the windows and doors closed and the air vent plugged. We ate, we watched movies and lots of CNN (since it was the U.S. presidential primary season and all), we played Guitar Hero and we tried our best to get a bit of work done while breathing the fumes in our stain-free environment. The taste and smell lasted in my mouth for days after the experiment—because of the persistent quality of PFCs, no doubt. At one point when I stepped outside the room (we left for only a few minutes at a time during the two days to get food from the kitchen, which we brought back into the room to eat on the sofa), I nearly fell flat on my face. The PFC coating on the bottom of my shoes made them so slippery I could hardly stand. At one point Rick's wife, Jennifer, came to visit and seemed quite appalled at what we were doing. She told us later that we both looked pale, completely red eyed and zoned out as we lounged with our feet up on the coffee table. She said we were like two stereotypical males watching Saturday night hockey, except with an alarming cloud of nasty chemicals around our heads.

Before the test began Rick and I had levels of four perfluorinated compounds in our blood that were similar to those measured in male Americans by the National Health and Nutrition Examination Survey (NHANES). My PFOA level was 2.8 nanograms per millilitre (ng/mL), and Rick's was 3.5 ng/mL compared to the NHANES mean value of 4.5 ng/mL.

Table 3. Bruce's and Rick's perfluorinated chemical blood levels compared to a sample of over 1,000 U.S. males as measured by the National Health and Nutrition Examination Survey (NHANES)[37]

PFC Type	Bruce's Results (ng/mL)	Rick's Results (ng/mL)	NHANES Geometric mean (ng/mL)
PFOA	2.8	3.5	4.5
PFOS	31.1	27.1	23.3
PFNA	1.2	1.1	1.1
PFHxS	1.9	2.7	2.2

As we sat in our room, chewing over our expectations for our experiments, we were convinced that if there was a test guaranteed to result in the most dramatic skyrocketing of our personal pollution levels, it was this one. With bisphenol A and the other things we were fooling around with, the actions we were undertaking were subtle. The chemicals were invisible, tasteless and (with the exception of the phthalates and personal-care products) odourless. The PFC experiment, by contrast, was dramatic. You could cut the off-gassing with a knife. We were sure we'd find hugely high levels of these chemicals in our bodies.

But it was not to be. As our blood testing showed, for the four chemicals of interest, including PFOA, there was no measurable increase in us over the two-day period. What had happened? We went back to Dr. Scott Mabury and his colleague Craig Butt for some answers.

"There could be two explanations for the lack of increase," said Butt. First, the high background exposure of PFCs and the relatively long half-life (about three to five years) of these compounds in our bodies likely makes it difficult to raise levels above the existing baseline. In addition, Butt pointed out, in experimental parlance we hadn't "controlled" for the product used by the carpet applicator. Although the guy told us he was using Teflon Advanced, which Mabury's lab has tested and confirmed contains the precursors to PFOA,[38] we never did see the original product container. This is in contrast to our other experiments, where we purchased the sources of pollution (such as tuna steaks, BPA baby bottles and antibacterial products) ourselves.

"I'm not surprised the experiment didn't work," concluded Butt. "Over the span of a few days it will not be possible to raise your blood levels above the background levels. Air exposure is probably an important exposure route, but it would probably take several weeks or more likely months of exposure to raise the levels above background."

Based on our experimental protocol, Butt did an approximate back-of-the-envelope calculation for Rick to illustrate his point.[39] His conclusion? Even if we had assumed unrealistic uptake and conversion rates, it would have been highly unlikely for Rick and me to see an increase in blood levels of PFOA over the two days of the experiment. Could a few weeks or months of exposure to the multiple sources commonly found in our homes and offices crank our levels? Yes. And it clearly does, judging from the levels of PFOA in all of us. But a few unpleasant days of trying to raise them deliberately did not work.

Too bad we hadn't figured this out before we spent two days sitting in that stinky room.

FOUR: THE NEW PCBs

[IN WHICH RICK FANS THE FLAME-RETARDANT FIRE]

Well it's been so long
And I've been putting out fire
with gasoline . . .
— DAVID BOWIE, "Cat People (Putting Out Fire)," 1982

THIS PART OF THE STORY started with some snuggly pyjamas. Really nice ones, covered with friendly penguins and dinosaurs, which my sister had brought back for our two young boys after some cross-border shopping at Carter's (a well-known chain of children's stores) in Buffalo, New York.

One day in late autumn as one of the coldest winters in living memory started to settle in, it was bedtime and I had our squirmy one-year-old, Owain, on his back. I started to put on the new one-piece pyjamas and stopped midway, the zipper halfway zipped, the upper part of the pyjamas bunched around his fat little baby belly.

"Sweetie!" I yelled down the stairs to my wife, Jennifer. "Have you seen this honkin' washing label on Owain's pyjamas?"

"What do you mean? No!" she yelled back. Owain tried to make a break for it. I grabbed him and pulled him back.

"Well, it's huge and talks about the material in these things being flame resistant."

By this point she was coming up the stairs. She knew where I was headed. "Does that mean you're not going to put them on now?"

she asked in a tone of voice typically reserved by the harried parents in our house for the 6 to 8 p.m. supper/bath/bedtime dash.

"Ummm . . . right. But I'll make you a deal, I'll phone the company and find out what flame retardants they put in these, and if it's not harmful I'll put them back on."

"Sure you will," she said, her face conveying her cumulative annoyance with my chemical skepticism, the non-stick pans relegated to basement storage, every plastic container upended and the symbols on the bottom scrutinized, every fine-print ingredient list on our shampoos examined.

But more on those pyjamas later.

Quest for Fire (Retardants)

In the climactic scene of the 1981 classic caveman flick, *Quest for Fire*, our neanderthal hero Naoh is taught how to make fire by rubbing sticks together by the Cro-Magnon Ivaka tribe. He is completely dumbfounded. No wonder. He and his friends have spent most of the movie wandering the primeval countryside, braving savage beasts and cannibalistic enemies on the assumption that fire is something that just exists. And if you lose it, you have to steal a spark from someone else. The fact that you can start it from scratch is a bit of a revelation. Once lit, though, a fire could do a lot of damage. How much do you want to bet that the very next thing our prehistoric ancestors learned how to do after starting a fire was how to quench an accidentally lit, out-of-control blaze?

We humans have been trying to master the chemistry of fire prevention for quite a while now. In ancient Egypt and China, vinegar and alum were painted on wood to increase their fire resistance.[1] During the siege of Piraeus by Sulla in 86 B.C.E., alum-soaked wood survived the fires of battle. In 17th-century Paris, flame-retardant treatments were pioneered for canvas, and in 1820 French King Louis XVIII commissioned the chemist Gay-Lussac to find better ways of protecting fabrics used in the theatre. Gay-Lussac is generally credited with being the first person to figure out the scientific basis of fire retardancy with his concoction of ammonium salts of sulphuric, hydrochloric and phosphoric acid. At about the same time, bromine was discovered in a French saltwater marsh, and our species' fire fighting was transformed forever.

Bromine—along with its pros and cons—is what we're really talking about. An element related to chlorine, fluorine and iodine (together called the "halogen" elements), it's a smelly, brownish liquid obtained from saltwater brine deposits. It turns out that bromine is pretty good at quenching fire. Usually, when something burns, the fragments interact with oxygen to keep the fire going. With the right kind of brominated mixture, the bromine atoms capture the burning fragments and prevent the combination with oxygen from happening. As a result the fire smoulders rather than spreading. Of the approximately 175 flame-retardant chemicals used at present, some of the most common—and most controversial—are "brominated."[2]

Because of this specific key ingredient, one of the unusual aspects of this family of chemicals is that its production is controlled by a very small number of companies with reliable access to bromine wells. The largest bromine reserve in the United States is located in Columbia and Union counties in Arkansas. China's bromine reserves are located in Shandong Province, and Israel's bromine reserves are contained in the waters of the Dead Sea. That's pretty much it for the world's current sources of bromine.

Also unusual for such a specific group of chemicals—and testament to the long-standing controversy swirling around them—is that brominated flame retardants (BFRs) are the subject of well-attended yearly international summit meetings. In June 2008 the Tenth Annual Workshop on BFRs was held in Victoria, British Columbia. Pretty much everyone who is anyone in the study and management of these chemicals was there. I decided to sign up to get a one-stop shop on the BFR story.

Victoria

Appropriately enough, given the workshop's focus on a chemical that comes from brine, the meeting was hosted at the Institute of Ocean Sciences (IOS) on the seashore just outside the city. On the surface the whole affair was polite and terribly arcane, including presentations with tongue-twister titles like "Hexabromocyclo-dodecane (HBCD) Alters the Expression of Genes Associated with Xenobiotic Receptor Activation and the Thyroid Hormone Pathway in Chicken Embryonic Hepatocytes." Dig a little deeper, however, and though couched in the tempered language of science, the workshop was riven with some highly controversial debates with serious, real-world implications.

The tone was set in the first morning's keynote address by Dr. Åke Bergman, an eminent Professor from Stockholm University. A bespectacled and somewhat grandfatherly figure for the assembled, his presentation was vast in its sweep—he gave a retrospective on the BFR question since the 1970s—and very strong in its conclusions. He reminded everyone that warnings about the health effects of BFRs were first raised decades ago, BFR contamination is now widespread throughout the world and it's finally time to ban some of the most commonly used of these compounds.

When I caught up with Dr. Bergman, he repeated his conclusion with the exasperation of someone who's been involved in a conversation that's gone on way too long: "I wish we could decide this afternoon to get rid of some of these flame retardants, because if we don't it will just cost a lot of work, a lot of meetings, for the next 10 to 15 years. We have to get rid of the additive brominated compounds that are lipophilic [meaning those that accumulate in our fat tissues]." I asked him why, and he talked about the damaging health effects of the chemicals and the strange logic of the flame-retardant industry—the perceived need to hose down everything with the stuff as opposed to dealing with the underlying problem: "I was travelling with our [Swedish] Deputy Minister for the Environment to the U.S. in 1999. We were in Washington, D.C., for a number of meetings. And we met with the CPSC [the federal Consumer Product Safety Commission]. The fellow we met with said, 'Of course, we need flame retardants on furniture, because kids are playing with matches.' I mean they were seriously saying that we need the flame retardants in furniture so the kids can go on playing with matches. I will never forget it."

Bergman pointed out that BFRs were products in perpetual search of a new use and that their application as flame retardants emerged with the demise of leaded gasoline. In the 1920s tetraethyl lead was invented as a gasoline additive, but it also left a corrosive byproduct in the engine. The solution hit upon at that time was to add a chemical called ethylene dibromide (EDB) to the mix. At this time gasoline additives accounted for about three-quarters of the bromine consumption in the United States.[3]

When leaded gasoline began to be phased out in the U.S. in the 1960s, bromine companies needed to dream up new applications for their product, and fast. Great Lakes Chemical Corporation—then the largest supplier of bromine products—decided to use EDB domestically as a pesticide. This plan ran aground, however, when in 1983 the U.S. Environmental Protection Agency issued an "emergency suspension" of all agricultural uses of EDB—the most restrictive measure the EPA can take under the law—because of evidence that EDB was a carcinogen and a mutagen and was contaminating groundwater supplies in a number of states. In a foreshadowing of bromine industry denials to come, Emerson Kampen, Great Lakes Chemical Corporation's President at the time, blamed the press. "It was the media that created the problem," he told reporters. "A great product has been taken off the market."[4]

But another use for bromine grew in the twilight of leaded gasoline: It began to be produced and marketed as a flame retardant. Great Lakes Chemical built several new flame-retardant plants in the early 1970s, and production of BFRs has been increasing ever since. At present more bromine goes into BFRs than any other application—about 40 per cent of global production. Dr. Tom Webster, a bearded and affable Professor at the Boston University School of Public Health and a senior BFR researcher, was also a participant at the Victoria workshop. I sat with him in the corridor outside the conference room and asked him how the BFR debate has changed over the years. "As Åke mentioned a little bit [in his keynote]," he told me, "specific events were very important. There was the Tris-BP controversy of the 1970s over flame retardants in children's pyjamas and whether they were mutagens or not. It was huge. I mean really huge. The next thing that happened was not that long after, and it was this business of polybrominated biphenyls in Michigan. The cows accidentally got it in their feed, and it caused this huge agricultural disaster."

"Threat to children and animals" was, in a nutshell, the image of BFRs that first hit the TV sets of consumers in the 1970s, interrupting their viewing of *M*A*S*H* and *The Brady Bunch*. Not a very auspicious introduction and a preview of the ongoing hullabaloo that still surrounds these chemicals.

Mutagenic Pyjamas

The story of Tris (2,3–dibromopropyl phosphate) highlights the tight relationship between the rise of BFRs and increasingly stringent flammability regulations that governments began to adopt in the 1970s. In the U.S., the *Flammable Fabrics Act* was first passed in the 1950s to regulate the manufacture of cool (and combustible) "High School Musical"–style clothing such as furry, pink brushed-rayon sweaters. The legislation was amended in the late 1960s to allow standards to be set for many additional consumer products.

In 1973, for the first time, the U.S. Department of Commerce set mandatory fire-resistance standards for children's nighties and pyjamas. Up to that point kids' PJs had mostly been made of soft cotton. Tris-BP quickly became the favoured chemical treatment, and because it was difficult to use with cotton, sparked the transition to polyester, the fabric most often used in kids' pyjamas to this day. Dollops of Tris-BP totalling about 5 per cent of fabric weight were layered onto the pyjamas of about 50 million U.S. children between 1973 and 1977. As the *New York Times* pointed out at the time, "the complicated tale of Tris . . . is a classic tale of good intentions—and of the sad truth of the axiom that, too often, the road to Hell is paved with just such generous or compassionate impulses, at least for the Federal Government."[5]

Though there was some early evidence of the new fire safety standards somewhat reducing the number of infants killed from their pyjamas igniting, evidence started to quickly mount that Tris-BP was a mutagen and a carcinogen. As a result, in early 1976 the Environmental Defense Fund (EDF) made big headlines when it petitioned the Consumer Product Safety Commission for action. Because the evidence linking Tris-BP to cancer was still somewhat equivocal, the EDF petition simply asked for Tris-BP garments to be labelled with a tag that said "Contains the flame retardant Tris. Should be washed at least three times prior to wearing."

The companies that manufactured Tris-BP pooh-poohed the petition, one of their spokespersons saying at the time that potential hazards from the chemical were extremely minimal.[6] The target of the petition, Richard Simpson, Chairman of the CPSC, said in an address to the American Apparel Manufacturers Association that "based on what I've examined . . . I would doubt that there is a problem. There is a great leap from the tests [cited in the petition] to a conclusion that the chemical is a carcinogen. And I am skeptical when a petition which raises the spectre of cancer suggests the remedy of a label asking people to wash the clothes three times before wearing. If there is a real problem, there should be a ban, not a label."[7]

Simpson had spoken too soon. In February 1977 the EDF obtained yet more evidence from National Cancer Institute testing that it claimed showed Tris-BP was a "potent" cause of cancer (one hundred times more powerful than the carcinogens in cigarette smoke) and that the chemical could be absorbed by children through the skin or by "mouthing" Tris-BP-treated children's clothing. The fund filed yet another petition—this time for a complete ban—with the CPSC.

Under pressure the industry caved. Sander Allen, a spokesman for the major manufacturer of Tris-BP, the Velsicol Chemical Corporation of Chicago, said in response to the fund's petition that the company did not agree that Tris-BP caused cancer but had discontinued making it for garments, because the safety testing necessary to assuage growing public concern was too costly.[8] Two things leap out from this quote that should spark a major feeling of *déjà vu* for intrepid readers of this book. First, apparently the industry was not required to do proper safety testing of Tris-BP *prior* to marketing its chemical, and second, even in advance of government action, the public debate sparked by the Environmental Defense Fund's first labelling petition had caused demand for Tris-BP-treated garments to collapse. Between 1976 and 1977 Tris-BP sleepwear went from about 60 to 70 per cent of the U.S. market to only 20 per cent.

With the heavy lifting completed by the EDF and the way cleared for its decision, in April 1977, exactly a year after Richard Simpson's dismissive comment, the CPSC acted on the National Cancer Institute's testing and banned the treatment of garments with Tris-BP. Not a moment too soon: A study published shortly afterwards in the journal *Science* actually found the chemical in the urine of children who were wearing or who had worn Tris-BP-treated sleepwear.[9]

Almost overnight an estimated 20 million garments in retail inventories were pulled from the shelves. Because of Tris-BP's toxic properties, the government prohibited the disposal of these pyjamas: They could only be buried or burned or used as industrial wiping clothes. What to do with this cancer-causing mountain of material? The answer quickly became evident as advertisements started popping up in the classified pages of publications like *Women's Wear Daily*: "TRIS-TRIS-TRIS . . . We will buy any fabric containing Tris-BP."[10] By some estimates, millions of Tris-BP-treated pyjamas were shipped quietly out of the U.S. to Europe and other parts unknown between the time Tris-BP was banned in 1977 and June 1978, when the CPSC also stopped Tris-BP exports.[11]

Propelled by the memory of the Tris-BP controversy, public demand for more comfortable pyjamas, made of natural fabric, grew throughout the next few decades, and in 1999 the CPSC finally relaxed its regulations on the flame retardants that needed to be added to PJs. At present less than 1 per cent of children's sleepwear is treated with flame retardants, although as we will see later in this chapter, the word "treated" as used by government and industry has a narrower definition than would be assumed by the average person.

As the Tris-BP fiasco was playing itself out over the U.S. airwaves, another geographically specific—though no less horrifying—BFR scandal was set to explode: the contamination of much of Michigan by polybrominated biphenyls (PBBs).

Cattlegate

One of the worst chemical disasters in U.S. history started on a single farm.

Rick Halbert knew there was something wrong with his dairy cows. His four hundred animals in southwestern Michigan were becoming increasingly unhealthy; they had decreased appetites and milk production, and they were developing really weird symptoms: hematomas and abscesses, abnormal hoof growth, hair matting and loss, and severe reproductive abnormalities. In the autumn of 1973, after his veterinarian was unable to diagnose any disease, Halbert suspected problems in his recent order of high-protein feed pellets, supplied by Michigan Farm Bureau Services, the state's largest feed distributor. Luckily for the people of Michigan, Halbert was no ordinary farmer. Before returning to the family business, he had completed a Master's degree in Chemical Engineering and had worked three years for the Dow Chemical Company. After repeated lack of response from Farm Bureau Services and the Michigan Department of Agriculture, Halbert spent five thousand dollars to conduct his own testing of the feed.

Though chemical contamination can have major effects, it often results from minuscule amounts of product. Sensitive equipment is needed to detect organic compounds such as those Halbert suspected were present. The more common chemical contaminants that were first considered—such as dieldrin, DDT (both of which pesticides were then in use) and polychlorinated biphenyls, or PCBs (of which PBBs are a very close cousin)—show up as early-emerging "peaks" on the readout produced by a scientific instrument called a gas chromatograph. Because of its extreme stability, PBB shows up as a late-emerging peak. Nothing was evident in the analyses of Halbert's feed until one day in January 1974, when the researchers running the chromatograph forgot to turn the machine off during their lunch break. When they returned, a remarkable and unfamiliar reading had appeared.[12]

Halbert passed this result to a scientist at the U.S. Department of Agriculture who, by coincidence, recognized it as a compound that he had been working with: PBB, a flame retardant produced by the Michigan Chemical Corporation for use in moulded plastic parts, such as the cases of televisions, typewriters and business machines. (Interestingly, Michigan Chemical was shortly thereafter bought up by the Velsicol Chemical Corporation, which was about to be embroiled in the Tris-BP scandal—see above.) The company also sold magnesium oxide to Farm Bureau Services, which then added this supplement to dairy feed to increase cows' milk production. There had been some sort of a mix-up at the company's plant.

When first informed of the situation in April 1974, Michigan Chemical rejected the idea that Firemaster, its PBB product, could have been substituted for Nutrimaster, its magnesium oxide

(MgO) product.[13] It was quickly proven wrong. As the state and federal governments moved in to investigate, it became clear that the mistake happened when the company ran out of preprinted bags and employees hand-lettered the two similar trade names on identical plain brown bags. After a bit of rough handling, and smudging of labels, between 225 and 450 kilograms of PBB were accidentally shipped to the Farm Bureau Services mill and incorporated into animal feed.

About nine months after the chemical first entered the Michigan food chain and Rick Halbert's cattle, the source was finally identified. It would be another year and a half before all the contaminated livestock and poultry were tracked down. By this point the contamination was widespread: Several thousand farm families and their neighbours had consumed poisoned meat, eggs and milk, and the general public in Michigan had been exposed to a wide array of PBB-contaminated products. The Michigan Long-Term PBB study has tracked the health of Michigan residents since the incident, and its results point to a potential link between high PBB exposure and an increased risk of cancers of the breast and the digestive system, lymphoma, elevated rates of spontaneous abortion and menstrual complications.[14]

By the end of 1975, about 28,000 cattle, 5,920 pigs and 1.5 million chickens had been destroyed. About 785 tonnes of contaminated animal feed, 8,137 kilograms of cheese, 1,192 kilograms of butter, 15,422 kilograms of milk products and nearly 5 million eggs were buried in huge pits throughout Michigan. Estimates of the total costs for statewide decontamination reached the hundreds of millions of U.S. dollars. And perhaps most shockingly, five years after the incident, about 97 per cent of state residents still had measurable levels of the chemical in their bodies.[15] No surprise that one of the most popular Michigan bumper stickers at the time read "PBB-Cattlegate Bigger Than Watergate."

In the end many investigations and much finger pointing ensued, and in 1982 the U.S. government, the State of Michigan and Velsicol Chemical announced a consent judgment of $38.5 million to settle clean-up costs associated with PBB and other contamination from Michigan Chemical Corporation.

As the 1970s drew to a close, a number of commentators expressed the hope that the newly adopted *U.S. Toxic Substances Control Act* (1976) would prevent Tris-BP and PBB-type disasters in the future.[16] But it was not to be. As Sonya Lunder, a scientist at the Environmental Working Group, so eloquently put it, "When we look at the history of fire retardants, we can assume that public health protection arrives late if at all."[17]

Yusho

PCBs, mentioned a few times already in this book, are familiar to many people, and with good reason. Short for polychlorinated biphenyls, PCBs, along with the pesticide DDT, are perhaps the most infamous of environmental contaminants. Manufactured for industrial applications including plasticizers, fluids in electric capacitors and hydraulic oils, PCBs were first detected in the environment in 1996—in the bodies of white-tailed sea eagles. Soon, scientists were measuring PCB levels in unlikely places all over the world, and this family of chemicals very quickly began to exhibit, in the words of the understated Dr. Åke Bergman, "very obvious toxic effects."

Seemingly out of nowhere in the summer of 1968, a terrible disaster occurred in the western part of Japan that dramatically illustrated the dangers of PCBs. PCBs leaked from the heat exchanger at a company that manufactured rice-bran cooking oil and contaminated some cans of oil that were then purchased by consumers. About 1,800 people were affected by what became known as "Yusho" (oil disease). Many fell gravely ill, babies were stillborn and about three hundred people died in the ensuing years from the poisoning. A whole host of very graphic symptoms were developed by sufferers, including angry sores on their faces and bodies (called "chloracne"); dark skin discoloration (even on newborn babies born to afflicted mothers); enlargement of, and hyper-secretion from, glands around the eyes; and respiratory and neurological problems.[18]

A very similar incident (causing an illness known as "Yu-Cheng") occurred in Taiwan in 1979. And the combination of the growing evidence of widespread environmental contamination by PCBs and the horrible illness and fatalities stemming from Yusho and Yu-Cheng moved governments the world over to act in unprecedented ways. Within the next few years, many nations had banned the production, and most uses, of PCBs. PCBs remain the only chemical specifically banned by a vote of the U.S. Congress (in an amendment to the U.S. *Toxic Substances Control Act*). And the international Stockholm Convention on Persistent Organic Pollutants brought a complete end to PCBs in 2001.

The banning of PCBs is one of the greatest environmental and human health success stories. Yes, the global environment remains contaminated with them three decades after they were banned, and most humans tested throughout the world have detectable levels in their bodies. But PCB levels are going down. I know this personally. As part of Environmental Defence's ongoing Toxic Nation campaign, I've been tested for these chemicals. I have nine kinds of PCBs at detectable levels in my body. A major drag, but I represent a point on an improving curve: The number of PCB compounds in my body, and the levels at which these substances were measured, are somewhat lower than in the study participants who are older than I am and much higher than in children who were tested.

Figure 3. The number of PCBs detected in Rick and the median number detected among Canadians tested in the Toxic Nation studies

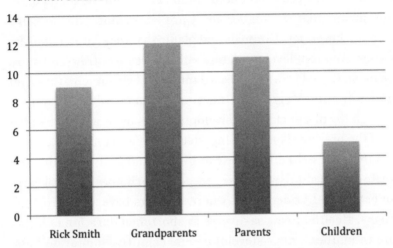

Another chemistry lesson is necessary at this point in our story. PCBs are a member of a family of chemicals called poly-halogenated POPs. Translation: They are persistent organic pollutants (POPs) containing many halogen (chlorine, bromine, fluorine or iodine) atoms. These chemicals have long half-lives in the environment and in the bodies of animals—about two to ten years. Other members of this chemical family include various chlorinated and brominated compounds such as PBBs (of Michigan cow fame, now phased out around the world) and poly-brominated diphenyl ethers (PBDEs)—currently among the most common flame retardants. The PCBs and PBDEs and others—let's call them subfamilies—each make up a group of individual chemicals called "congeners." Each congener shares the subfamily's chemical backbone but has different numbers and positions of halogen atoms attached.

Persistent organic pollutants have three chemical characteristics that make them intrinsically hazardous: They are stable (persistent), they are stored in fat tissue for long periods of time (that is, they are "lipophilic") and they have the potential to act as endocrine (or hormone) disruptors. The stability and lipophilic nature of POPs causes them to "biomagnify" up the food chain. This means that the higher-level predators store all the POPs from the lower-level animals and plants that they eat—in their fat tissues. And top-level predators—like us—concentrate and store the most. Once POPs are released into the environment, they find their way into pregnant or nursing mothers, where they pass through the placenta to the developing fetus or concentrate in the fat of the breastmilk and are ingested by the nursing infant.[19]

Some POPs—like PBDEs—can bind to our cell receptors and create effects similar to those caused by hormones. Several carry out estrogen-like activities, whereas others have antiestrogenic effects. Health effects as diverse as shortened duration of lactation in mothers, neurodevelopmental cognitive-motor deficits, intellectual impairment in children and greater risk of cancer have been attributed to polyhalogenated POPs.[20]

In short, PBDEs and PCBs are so similar that some scientists are increasingly referring to the former as "the new PCBs." But as we shall see, unlike the case with PCBs, the challenge of global PBDE contamination is a long, long way from being solved.[21]

Mother's Milk

As Dr. Deborah Rice tells it, it was the similarity of PBDEs to PCBs that first caught the interest of the scientific community. Rice, a scientist at the Maine Center for Disease Control and Prevention, and a well-known expert on brominated flame retardants, spoke with me on the phone from her office in Augusta. "Analytical chemists looked at the structure of PBDEs and they said, 'Wow! This looks a lot like PCBs and some of the other chemicals that are out there that we know are persistent, that we know are bioaccumulative, and gee I wonder if that's what PBDEs are doing out there in the environment.'

"We came to recognize that just like PCBs, these chemicals are indeed persistent," she said. "They bioaccumulate through food chains all over the world and of course we're terminal predators [at the top of the food chain], so we get a concentrated dose. When you take PBDEs into the laboratory and look at them for endocrine disruption, for disruption of thyroid hormones and for developmental neurotoxicity, they have the same effects, the effects you'd expect them to have based on their similarity to PCBs.

"We never seem to learn!" Rice exclaimed. "To me it's really a very sad tale that as we were banning PCBs in the late 1970s, we were putting PBDEs and other flame retardants into the environment."

In a prominent 1977 *Science* article about the Tris-BP controversy, Arlene Blum and Bruce Ames actually warned of the possibility of more widespread global pollution by flame retardants. This last sentence of their article encapsulates both their impatience and the growing challenge in dealing with an avalanche of synthetic chemicals: "While waiting for the effects of the large-scale human exposure to the halogenated carcinogens—polychlorinated biphenyls (PCBs), vinyl chloride, Strobane-toxaphene, aldrin-dieldrin, DDT, trichloroethylene, dibromochloropropane, chloroform, ethylene dibromide, Kepone-mirex, heptachlor-chlordane, pentachloronitrobenzene, and so forth—we might think about the avoidance of a similar situation with flame retardants."[22]

But in much the same way that the world started to wake up to the dangers of PCBs only in the late 1960s, after they started to be found in wildlife, Tom Webster (of the Boston University School of Public Health) dates the first awareness of the dangers of PBDEs to a very specific moment: "With PBDEs they were pretty much under the radar until about 1998 and that's when this group in Sweden did [a] . . . retrospective study of breast-milk. And what they found is that the levels of PCBs and dioxins had started coming down from about 1970 to 1998 and levels of PBDEs were going up. And that totally got everyone's attention. Because PBDE levels are going up exponentially and they're basically cousins of PCBs."

The Swedish results were a bombshell. Humans are at the top of the food chain for sure, but nursing babies are at the very top. Mother's milk concentrates lipophilic pollutants like PCBs and PBDEs and passes them on to the next generation (though it's important to point out that the health benefits of nursing still outweigh the downsides). Using banked breastmilk from thousands of Swedish women sampled between 1972 and 1997, the researchers were able to show diametrically different results for PCBs and some other contaminants and PBDEs. The level of PCBs in the milk banked in 1997 was 30 per cent of the level found in milk from 1972. In stark contrast the concentration of PBDEs in the milk increased exponentially between 1972 and 1997, doubling every five years.

These results were widely publicized and so astonishing that they moved a number of key individuals who still work in the field to begin new research programs on PBDEs. Tom Webster, who used to do work on dioxin, is one. Dr. Heather Stapleton, a young, up-and-coming academic at Duke University, who frequently collaborates with Webster, is another. She was a graduate student at the time the Swedish study was released, looking at PCBs and organochlorine pesticides in Lake Michigan. When the Swedish research hit, she immediately started testing for PBDEs in her Lake Michigan samples. More on her work a bit later.

With each passing year the massive dimensions of the problem become clearer. It was overwhelming listening to the presentations in Victoria and walking down the aisles of posters illustrating the newest science in this area. The titles of the studies of the assembled international scientists tell the tale:

- "Association between PBDE Exposure and Preterm Birth." (Good evidence from a big sampling of U.S. moms that PBDE pollution causes premature labour.)
- "PBDEs in the St. Lawrence: New Contaminants to Be Monitored." (Over the last ten years, concentrations

have increased fivefold around Quebec City. Trends
around the world are similar.)
- "PBDEs in Harbor Seals from British Columbia,
  Canada, and Washington State, U.S.A.: An Emerging
  Threat." (Levels are increasing exponentially and will
  surpass the levels of PCBs by 2010.)
- "Levels and Patterns of PBDEs Measured in Human
  Tissues from Four Continents and in Food and
  Environmental Samples from Selected Countries."
  (They're everywhere.)

You get the picture.

Dust to Dust

One of the basic questions that PBDE researchers continue to
puzzle over is how PBDEs get into people. It seems to be quite different from other persistent, bioaccumulative toxins that enter the
human body, mostly through food. In fact, for a number of years,
the levels of PBDEs being measured in Americans didn't make any
sense to scientists at all. Some people had hugely elevated levels—
100–500 per cent above the national average. With food-borne
pollution, levels across populations were more similar because
everyone eats food. There was obviously some other unidentified
source of PBDEs in everyday life that could cause marked differ-
ences between individuals and households. But what was it?
Heather Stapleton was one of the first to solve this riddle.

She recalls that shortly after PBDEs were noticed by the scientific community, everyone started measuring levels in aquatic
environments, soils and sediments. And yet PBDEs are used in
objects like sofas, rugs and television sets. "I remember thinking
to myself that we were measuring for these things in external environments, when they're actually found in internal environments."
So Stapleton took a sample of house dust she had in her lab, which
she was testing for pesticides and lead, and analyzed it for PBDEs

instead. She was shocked at the results. Levels of flame retardants were much higher than she had expected.

Next, she measured dust samples from 16 homes in the Washington, D.C., area. Again, very high levels of PBDEs. Stapleton published this research at about the same time that PBDE levels in Canadian homes were being measured, and as she says, "When we presented this, it really opened people's eyes. It made sense. It all fell into place. It was a different paradigm about how we think about the sources of, and exposure to, these compounds." It turns out that PBDEs leach out of the products they are put into: the squishy foam in a sofa, the padding in a mattress and the back of a TV set. They waft into the air in our houses and offices and cars and sailboats and settle to the ground as dust.

PBDEs and Rug Rats

Stapleton's research points to the extreme risk of children being exposed to PBDE pollution. Because they're closer to the floor, because they play in places where dust lurks (such as under the bed—one of my sons' favourite hangouts), children are uniquely vulnerable to exposure to this sort of indoor pollution. And developing research—including our Toxic Nation study—on comparative levels in different ages of people bears this out: Younger people have higher levels of PBDEs than older people do. PBDEs were not around when my grandparents were born. They spent much of their life unexposed to these toxins. In my childhood PBDEs existed, but they were not as common as they are now. My two young boys are wallowing in the stuff every day and will bump up against this pollution for the rest of their lives. Unlike the PCB success, PBDEs—and some other newer pollutants—are a legacy we are leaving to our children.

Heather Stapleton has continued her research and presented a study at the Victoria meeting on two new kinds of BFRs that she recently found in household dust: 2 (ethylhexyl) 2,3,4,5-tetra-bromobenzoate (TBB) and bis (2-ethylhexyl) tetrabromophthalate

(TBPH). These chemicals are the basis of a new flame-retardant mixture called Firemaster 550, manufactured by Chemtura Corporation (a new company formed by the amalgamation of Great Lakes Chemical and Crompton Corporation in 2005). Stapleton was told by chemical and furniture manufacturing companies that "they would be very surprised" if she found the new BFRs in household dust. Well, she did. Three to four years after Firemaster 550 first appeared on the market as the preferred flame retardant for the polyurethane foam used in many North American sofas, its chemical ingredients were present at detectable levels in household dust.

Who knew that dust bunnies could be such dangerous beasts?

With all this evidence of contamination and harm, with all these similarities to PCBs and the discovery that PBDEs are now polluting the most private recesses of our homes, the million-dollar question is obviously "Why haven't PBDEs been banned yet?"

Deborah Rice (of the Maine Center for Disease Control and Prevention) thinks it's because there's been no Yusho-style catastrophic accident with PBDEs so far. Whatever pollution is happening is slow. It's quieter and more insidious than PCBs. Nobody's yet died in a graphic illustration of what happens when you get a huge dose of the stuff, allowing people to extrapolate and realize the harm being done to themselves, over time, from much smaller doses. To be crass about it, the bromine industry has not been brought to task, because no one has yet been able to produce a dead body linked to PBDE poisoning.

"You're Human!"

I went to the Victoria meeting armed with my PBDE blood test results. Unlike other chemicals to which I'd subjected myself for this book, I didn't try to manipulate levels of PBDEs in my own body. I just offered up ten vials of blood for a one-time test and called it a day.

The reason for this decision was that after speaking with a few experts, the consensus seemed to be that PBDEs are so prevalent and have such long half-lives (measuring in years) that it would be impossible to demonstrably affect levels in me within a few days. Actually, that's not entirely correct. I probably could have jacked up my levels of PBDEs very quickly, but in order to do so, I would have had to stray into outlandish activities and thereby violate the one immutable law that Bruce and I agreed would define our experiments—that we needed to stick with "everyday" activities.

Short of jumping up and down on an old decaying couch for hours on end, filling a room with PBDE-laden dust and aspirating it with vacuum-like zeal or sitting down and eating some dust bunnies for breakfast, the process of affecting my PBDE levels seemed likely to take weeks or months, not days. (Interestingly, however, in Victoria, Åke Bergman presented the results of a brand-new study in which eight Swedish travellers showed significantly increased PBDE levels after taking long-haul flights overseas and returning home a number of days later. Aircraft, with all their upholstery and foam insulation and closed air systems, are extremely high in PBDE contamination, so perhaps under other specific conditions like this, PBDE levels can be affected relatively quickly.)

Certainly, levels of PBDEs can be changed over time. For those of us interested in decreasing the levels in our bodies, all is not lost. Levels of some specific flame retardants are now starting to decrease in human milk in Sweden, for example, after that government took action to ban them. Avoiding PBDE-containing products will certainly make a difference. It's just that we didn't have the financial resources to mount an experiment to demonstrate this over a period of years.

In my own body it turns out I have eight detectable PBDEs. "You're human!" joked Coreen Hamilton, Director of Research at Axys Analytical Services, the lab that evaluated many of our blood and urine samples, as she reviewed my results. "Everyone's got similar levels. These look pretty typical."

Table 4. Rick's PBDE results are shown in comparison to the National Health and Nutrition Examination Survey (NHANES) results. The NHANES data is based on blood samples collected from U.S. citizens to be analyzed for PBDE levels.

BDE Type	My Results (ng/g lipid)	NHANES Geometric mean (ng/g lipid)	NHANES Median mean (ng/g lipid)	NHANES 25th percentile (ng/g lipid)
BDE 15	0.1	NA	NA	NA
BDE 28	0.4	1.2	1.1	<LOD
BDE 47	8.1	20.5	19.2	9.3
BDE 49	0.1	NA	NA	NA
BDE 99	1.3	5.0	<LOD	<LOD
BDE 100	1.2	3.9	3.6	1.6
BDE 153	4.1	5.7	4.8	2.4
BDE 154	0.1	*	NA	NA

* Not reported, 75th percentile is 0.8.
"NA" means not measured by NHANES.
"LOD" means result less than level of detection of analysis.

I dug a bit deeper with Tom Webster (of the Boston University School of Public Health), who compared my results to the huge National Health and Nutrition Examination Survey (NHANES) study—a regular survey of levels of environmental contaminants in the U.S. population by the Centers for Disease Control.[23] Webster contrasted my results with the geometric mean, the 50th percentile and the 25th percentile of the 2,062 NHANES study participants.

"As you can see, for all reported congeners, your values are less than the geometric mean and 50th percentile. For BDE 47 and BDE 100, you're also less than the 25th percentile. For BDE 153 you're between the 25th and 50th percentiles. So overall, I'd say that your PBDE[24] concentrations are less than the typical U.S. resident. There have been some suggestions that Canadians have lower levels than Americans, on average, but until we have a representative sample of Canadians, we won't know for sure."

I couldn't stop thinking about Coreen's offhand comment. How strange and disturbing that along with the possession of two eyebrows and ten toes, the defining characteristics of most humans on earth now includes measurable levels of brominated diphenyl ether #153.

If they want to, future archeologists will be able to easily define toxic eras in human history from the telltale levels of these potent, globally sprinkled chemicals in our race's desiccated remains. The classifications might go something like this: Era PCB (1950 to 2030); Epoch PBDE (1980 to 2075); Age of Firemaster 550 (2005 to ?). Like the rings of a tree or the layers in sedimentary rock.

The Global Bromine Oligopoly

Deborah Rice thinks there's a second big difference that explains the relatively quick ban on PCBs as opposed to the excruciatingly slow progress on PBDEs. And she's better positioned than anyone to have an informed opinion on this. Quite simply, industry is fighting harder to keep PBDEs than it did for PCBs. The bromine industry is organized.

As the *Chemical Marketing Reporter* put it, "the global bromine industry is essentially an oligopoly controlled by Albemarle, Great Lakes [now Chemtura Corporation], and the Dead Sea Bromine Group [now Israel Chemicals]."[25] In 1997 these companies banded together with the Tosoh Corporation to form the Bromine Science and Environmental Forum (BSEF) to lobby for their interests, and together, BSEF members control over 80 per cent of the global production of brominated flame retardants. A sort of OPEC of BFR.

These companies have obviously been successful at advocating for their interests. Market demand for flame retardants increased from 372 million kilograms in 1996 to 450 million kilograms in 2006. By 2011 demand is expected to reach just over one billion dollars.[26] PBDEs are now found in a huge number of common consumer items, with the majority being used in the ever-increasing panoply of electronic devices, gizmos and googas that fill our lives.

Unlike other industrial groups I've dealt with over the years, the bromine barons push their products through lobbying, in many cases, for *more* government regulation. Usually, industries want to avoid regulation like the plague and engage in huge lobbying offensives to derail it entirely or to make sure it's so toothless that they can safely ignore it. And, of course, as we'll see in a moment, BSEF members do their fair share of this. But at the same time that bromine companies are trying to convince governments to leave their affairs as unregulated as possible, they push relentlessly for tighter and tighter regulations on the business of others through increasingly stringent flammability standards on manufactured goods. As we've seen going all the way back to the use of Tris-BP in pyjamas, fire-prevention standards mean darn good business for bromine companies. The higher the standards, the better. Not surprisingly, the BSEF is represented on key committees in major jurisdictions that actually make the decisions on new flammability standards.

I met Joel Tenney, the North American Advocacy Director for Israel Chemicals, and the BSEF's Canadian lobbyist, Chris Benedetti, at the Victoria meeting on brominated flame retardants. Both nice, low-key guys. Both terribly "on-message" when it came to the benefits of the industry's brominated products. Listening to them you'd think that BSEF members are really in the business of public service, just doing their part to save lives and property and combat accidental fires (and, of course, they can quote chapter and verse on the number of accidental fires specifically avoided due to the wonders of BFRs).[27] The seeming reasonableness of the industry's "Elmer the Safety Elephant"–style message and its spokespeople belies the ruthless way that the industry has dealt with its critics over the years and sought to stall any progress on bromine regulation.

The PCB Playbook
When PCBs were banned industry's attempts to forestall the inevitable lasted only a few years. The evidence of global contamination and human health problems linked to PCBs, along with

numerous PCB-spill accidents throughout Europe and the United States, made it clear pretty quickly that the jig was up. An insight into the industry's thinking can be gleaned from a fascinating internal 1969 discussion paper from Monsanto—the only PCB manufacturer in the U.S. at the time. Entitled "PCB Environmental Pollution Abatement Plan," the document acknowledges that "PCBs are a worldwide ecological problem" and sets out three possible courses of action, noting the pros and cons of each. Here's an excerpt from the Monsanto "playbook":

> 1) Do Nothing—"We would most likely be forced out of this business. Other product areas would be adversely affected. We would project an image as an irresponsible member of the business world."
> 2) Discontinue Manufacture of All PCBs—"Although we all realize this could be an eventually [sic], unfortunately the solution is not this simple. . . . Financial loss could be considerable. . . . Competition would take advantage on all fronts. We would be admitting guilt by our actions."
> 3) Respond responsibly, admitting that there is growing evidence of environmental contamination by the higher chlorinated biphenyls and take action as new data is generated to correct the problem—"We could maximize the corporate image by publicizing this act. . . . Additionally we could gain precious time needed to develop new products and investigate further the lower chlorinated materials."[28]

Monsanto chose Option 3 (seem reasonable on the surface but do everything possible to delay) over Option 1 (seem obstreperous and patently unreasonable) or Option 2 (do the right thing). Similar approaches are taken by bromine companies today.

In 1970 Monsanto announced that it would no longer sell PCBs for use as a water-resistant plasticizer or as a hydraulic fluid but that it would continue to manufacture the chemical for use as a coolant in electrical transformers.[29] In spite of the unstoppable momentum behind a PCB ban, the company kept producing PCBs for almost six more years. It wasn't until early 1976 that Monsanto announced it was planning to phase out the manufacture of PCBs completely. Even then the company spokesman said he couldn't give an exact timetable for the phase-out.[30]

Fast-forward to Europe in the late 1990s in the wake of the Swedish breastmilk study. While protesting all the while that their products were safe, BSEF members started to soften their public line defending "Penta," one of the three commonly used PBDEs, but they retrenched around defending the other two PBDEs, "Octa" and "Deca." When the European Union and California proceeded to ban Penta and Octa in 2003, Great Lakes Chemical announced it would voluntarily phase out these two chemicals by 2005 but ramped up its defence of Deca and newer products like Firemaster 550. "Deca-BDE is the most widely used of the three and has been tested extensively by the National Academy of Sciences, the World Health Organization, and the EPA [the U.S. Environmental Protection Agency]," BSEF said at the time. "All have given Deca-BDE a clean bill of health."

Tris-BP and PBBs in the 1970s. Ethylene dibromide in the 1980s. Penta and Octa in the 1990s. Deca today. The bromine industry's use of the "PCB Playbook" has succeeded for almost 40 years.

Two Down, One to Go

Deca was the hot issue in Victoria.

Right from the first few minutes of Åke Bergman's presentation ("It has to go. The evidence to ban Deca is so convincing"), the scientific consensus was clear: Deca is emitted from many consumer products, it breaks down or "debrominates" in the environment into the more harmful Penta and Octa PBDE varieties and it is causing real health problems for wildlife and humans. The industry still denies all this and is fighting aggressively on three fronts—Europe, the United States and Canada—to head off a complete Deca ban.

On July 1, 2008, Europe finally banned Deca in electronics—a major blow for BSEF, given that electronics were slurping up the vast majority of the Deca produced. Canada has labelled Deca "toxic" under its national pollution law and is still considering whether to follow Europe's lead.

Meanwhile, in the U.S., things are going better for the industry. It's been successful, so far, in ensuring that Deca bans are restricted to the state level. To date, only Washington and Maine have instituted bans. And the industry has sent a very clear message that it will steamroll anyone who stands in its way. In 2007 the bromine industry lobby was successful in having Deborah Rice removed as the Chair of the Environmental Protection Agency's expert peer review panel charged with setting safe exposure levels for Deca. In a letter to the EPA, a lobbyist for the American Chemistry Council suggested that Rice—who works for the Maine Center for Disease Control and Prevention—should be turfed from the panel because she exhibited an "appearance of bias" when she represented her employer's position in favour of restricting Deca in her testimony before the Maine legislature. EPA buckled to the industry pressure not only by removing Rice from the panel but also by trying to cover up its actions by stripping any mention of her and her comments from the original published review documents. The altered document was reposted on the web with no indication that any changes had been made.

Rice's removal was outrageous for two reasons. The first was that it illustrates how seriously the EPA takes its role as an industry lapdog. The speed with which it acted on the industry's complaints is staggering. Another reason is that the EPA apparently sees no "appearance of bias" when it comes to people who are cozy with the chemical industry. In a small but telling study, the Environmental Working Group examined seven EPA panels established in 2007 that use non-EPA scientists to evaluate the agency's proposed safe exposure levels. EWG identified no fewer than 17 individuals who were employees of companies that were making the chemicals under review, scientists whose work was funded by industries with a financial stake in the panel outcome or scientists who had made incautious statements about the safety of the chemical in question.

Despite her obvious annoyance at being personally caught up in the industry's lobbying machinations, Deborah Rice thinks that the defenders of bromine have a "tin ear" when it comes to changing public attitudes toward chemicals and are losing ground because of their overly aggressive tactics. She mentions a few specific incidents that occurred during Maine's debate on Deca: "They couldn't have shot themselves in the foot any harder than they did. They came in completely heavy handed, running full-page ads in the paper and mischaracterizing the bill that was before the legislature. For weeks and weeks they had TV ads. It was purely scare tactics and treated folks up here like we were a bunch of idiotic country bumpkins."

This approach completely backfired, says Rice. "The legislators were really quite aggravated by the industry's tactics. It really stood out in Maine because we're a small state, we have a small population, we're a poor state, we're a rural state; we don't get this much attention about anything, ever."

Laurie Valeriano, the Policy Director at the Washington Toxics Coalition, tells a similar story of what happened when the bromine industry "descended" with its "aggressive, misleading and well-funded" campaign to head off Washington State's Deca ban. "They ran ads basically saying that we were going to compromise fire safety. But they didn't have the right spokespeople to pull it off. They didn't have the Fire Fighters, they didn't have the Fire Chiefs, they didn't have the Fire Marshalls to back up their claims that this was going to compromise fire safety in any way. They just lacked credibility."

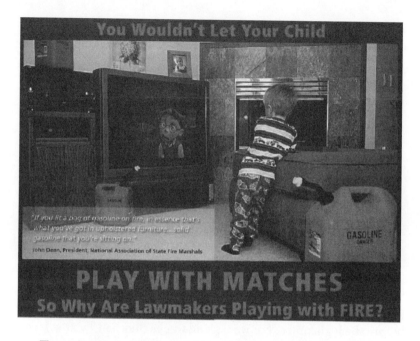

Flyers distributed by the bromine industry to Washington State
households to avert a state-wide ban on Deca

This lack of credibility wasn't helped, Valeriano says, when the
industry "brought in a poor gentleman to speak to legislators
who was badly burned in an airplane fire and yet our bill didn't
even cover airplane flame retardants." One of the turning points in
the debate was when "a paid employee of Ameribrom, a company
related to ICL [Israel Chemicals Limited, one of the largest brominated flame retardants manufacturers] and Albemarle, came and
testified to the legislature as representing the 'National Fallen Fire
Fighters Foundation.' He didn't disclose to the legislative committee that he was actually a paid employee of the bromine industry;
the legislators weren't happy that they were being hoodwinked and
all this wound up on the front page of the newspaper."

Money can't buy you love. Or in the case of the bromine industry in Maine and Washington, effective lobbying.

Putting Out Fire with Gasoline

My wife and I are addicted, at the moment, to the TV series *Mad Men*. As we watched the latest episode, I suddenly thought that the reality depicted in the show would be the kind of world the bromine industry would really enjoy. *Mad Men* follows the lives of a handful of hard-drinking, philandering advertising executives in New York City in the early 1960s. Everyone in the show smokes. At work, in boardrooms, in their offices, in the bathroom, in bed, in their cars. Watching it is really an assault on the senses. Such a world, with so much potential for house fires caused by lit cigarettes, is a bromine industry executive's dream. The more fires there are, the more loss of life, the more property damage, the more support there would appear to be for infusing every household item with brominated flame retardants.

One of the greatest challenges to the bromine industry's ongoing attempts to convince companies to flame-retard so many products has come from an unlikely source: the international movement to legislate self-extinguishing or "Reduced Ignition Propensity" (RIP—an unfortunate acronym) cigarettes. Canada was the first country in the world to require, in 2005, that cigarettes be made from special paper that has bands, or "speed bumps," to slow the burn. Europe is moving toward a similar continent-wide standard. And in the absence of a national approach, advocates in the United States are proceeding on a state-by-state basis. As of this writing 82 per cent of Americans were protected through state-level cigarette regulations. To the extent that smouldering cigarettes are the leading cause of fire deaths and this new kind of cigarette will, by some estimates, eliminate three-quarters of these fatalities, the bromine industry may have finally met a flammability standard from which it derives no benefit. Less risk of fire equals less traction for their arguments.

Even here, however, the bromine industry's intransigence is incredible. In a fascinating development it turns out that one of the current top lobbyists for the bromine industry, Peter Sparber, is a former top lobbyist for the tobacco industry. Sparber has spearheaded efforts to oppose the spread of self-extinguishing cigarettes for both his current and past employers (the cigarette companies wanted to avoid further regulation). Sparber has also been involved in efforts to petition the Consumer Product Safety Commission to require furniture manufacturers to make upholstered furniture so it would resist ignition by a lit cigarette or a small open flame, an elevated standard that can best be accomplished with brominated flame retardants.

Thankfully, it seems likely that the CPSC will resist this industry approach, I am told by Arlene Blum. In the late 1970s it was Blum's research that crystallized the problem of Tris-BP in sleepwear, and in the last couple of years, she has led the charge against flame retardants in other consumer items. She was prompted to do so after one of her cats was diagnosed with thyroid disease and was found to have high levels of the chemicals in its blood.

"The marketing plans of the fire-retardant industry rely on having candle or open-flame standards as much as possible," Blum says. "There's only one candle standard right now that exists, and that's the California furniture standard. Because of this standard, which has been around since the early 80s, all the furniture in California is supposed to resist an open flame and is heavily reliant on toxic flame retardants." Contrary to the bromine industry's rhetoric, other states that don't have such a standard have done just as well as California in reducing fire deaths since 1980. "The CPSC seems to get this," Blum continues. "So they've recently unveiled a new model that instead of fire-retarding foam, they are going to require fabric to just be smoulder resistant, not open-flame resistant. And with that model you can retard flammability without high levels of chemicals."

The industry's lobbying efforts won't be helped by the fact that a major recent study has found that because of their state's unnecessarily PBDE-dependent flammability standards, Californians have nearly twice the national average of PBDEs in their blood.[31]

A total of 36,553,215 people (according to the 2007 estimate of California's total population) are being poisoned by their impeccable upholstery.

Arlene Blum is positively excited about duking it out with BFR proponents in the murky world of national and international product standards. She is currently doing battle with little-known agencies such as the International Electrotechnical Commission, European Committee for Electrotechnical Standardization (CENELEC), Underwriters Laboratory and the Canadian Standards Association. If the bromine industry is successful in pushing open-flame standards in these venues, it would mean more mountains of toxic flame retardants coming into homes, schools, hospitals and businesses—wherever electronic equipment is found. As the bromine companies lose ground in open and transparent legislative debates, you can imagine that they may turn increasingly to peddling their wares in these more arcane and hidden ways. Blum's fight against BFRs has been going on for over 30 years. Like many aspects of the flame-retardant debate, it has the "Same as it ever was feeling" of that famous Talking Heads lyric.

Throughout my interviewing for this chapter, I asked everyone whether we're learning any lessons—as scientists, as government regulators, as a society. Or whether we're doomed to lurch from one flame retardant to another, substituting the harmful with the even worse. Opinions were mixed.

Some expert observers see nothing changing. Tom Webster was skeptical that lessons are being learned but felt that at least "the cycle is speeding up." That is, toxicity is being diagnosed faster and harmful chemicals are being removed from the system earlier than in the past.

I took heart from the fact that Åke Bergman, the veteran of the bunch, has started to think just over the past half year that changes are happening. Perhaps after the E.U.'s decision to ban Deca in electronics, the bromine industry is finally seeing the writing on the wall for PBDEs. This possibility would seem to be buttressed by a recent analysis from the international business research company, the Freedonia Group, which forecasts continued growth through 2011 in the use of flame retardants worldwide, but with a significant caveat: "Phosphorus-based flame retardants will grow at the fastest pace, driven by increasing trends toward non-halogenated products. However, brominated compounds will continue to lead the market in total value, as the regulatory climate in the US is unlikely to undergo dramatic changes in the near future. . . . Increased regulatory pressure on halogenated compounds is leading brominated flame retardant suppliers to diversify their product lines, a trend underscored by the forthcoming acquisition of Supresta by Israel Chemicals, which is expected to close in late summer 2007."[32]

The future may be flame retarded but perhaps not brominated.

Owain's Pyjamas

Jennifer was right, of course. I never did get around to phoning about the pyjamas in time to use them that winter. Our kids continued to wear their somewhat ratty hand-me-downs to bed. And it was just as well. After the PJs sat on my desk for eight months, I finally got tired of looking at them and phoned Carter's to ask some questions.

"There are no flame retardants in our 100 per cent polyester pyjamas," I was assured by the nice woman at the other end of the customer service line. "They're all natural. The polyester is naturally fire resistant." Not aware that polyester is "all-natural" anything, I asked if she could send me something in writing to confirm this. Within minutes (leading me to believe that perhaps they'd heard this question before), I received a short document through my email, emphasizing that Carter's products "are made of polyester which complies 100% with CPSC guidelines."

After doing a bit of research on what these CPSC guidelines are, I discovered that most polyester in sleepwear is now infused with a few different kinds of flame retardants. It's not painted on the surface as Tris-BP was (the CPSC calls this "treated") but rather bonded right into the fabric. Chemicals used in this way include halogenated hydrocarbons (chlorine and bromine), inorganic flame retardants (antimony oxides) and phosphate-based compounds. I sent an email back to Carter's, asking exactly what flame retardant is mixed into their polyester, and a week later I received this message from the Quality Department: "We rely on the natural flame resistant properties of polyester. When manufactured in a clean environment we meet all applicable state and federal regulations." That didn't really answer my question.

I left it there but sought advice from Duke University Professor Heather Stapleton—the researcher who'd discovered the Firemaster 550 in house dust after it was supposed to remain forever bound in the products into which it was inserted.

"Are you careful in your personal life to try and avoid PBDE-laced products?" I asked.

"I am when I can be," she replied. "For example, I don't like to have carpets in my home; I prefer hardwood floors. . . . IKEA has moved away from all halogenated flame retardants, so I try to buy furniture from IKEA. I also try to stay away from STAINMASTER-treated furniture, because I'm concerned about the perfluorinated chemicals [see Chapter 3] they put in the STAINMASTER. . . . I try to be very aware of what I'm buying for my own home."

She continued: "My fiancé and I are interested in having kids over the next couple of years, too, so I'm definitely interested in trying to reduce any kind of exposure to these chemicals. . . . It's just a shame that you want to go out and buy a TV, for example, and you have no idea if it's been treated with any flame retardants. The only way to scan it is with an expensive XRF instrument to see if it contains bromine or not, and not everybody has access to these instruments."

I told her about my questioning Carter's and asked whether, at the point when she had kids, she would have a preference for flame-retarded or non-flame-retarded PJs.

"Yeah, I would choose the non-flame-retarded ones without a doubt."

Good enough for me. We're sticking with cotton. And Owain's penguin PJs are still sitting on my desk.

FIVE: QUICKSILVER, SLOW DEATH
[IN WHICH BRUCE EATS MUCH TUNA]

I was eating tuna four times a week. I had crying spells,
low-grade depression, loss of memory, and brain fog, which is
where I would be talking to you and I would get disoriented.[1]

—DAPHNE ZUNIGA, ACTRESS

ALL SHE WAS DOING was eating "your average Hollywood stay-in-shape diet, a ton of fish and low carbs," actress Daphne Zuniga told *ABC News* in 2005. Perhaps best known for her starring role in the 1990s TV series *Melrose Place*, Zuniga recollected that she "would go out for sushi and think, 'Oh, great, at least we're not going for Italian, with all the oil and carbs.'"[2] Over time, however, she noticed unusual symptoms, including an itchy rash all over her body that landed her in the emergency room. She saw plenty of doctors, but nobody seemed to have a clue. It was only after reading a commonly quoted statistic from a U.S. Environmental Protection Agency study to the effect that one in six women of childbearing age has elevated mercury levels that she thought she should go for tests. Sure enough, her blood mercury levels were significantly over the safe level. She changed her diet, and the symptoms largely disappeared within six months.

In addition to potentially addressing the question of why so many of the stars interviewed in *People* magazine seem so distracted (Answer: They're mercury-addled), Zuniga's cautionary tale also underlines mercury's credentials as one of the most potent neurotoxins known. It specializes in attacking the brain. That mercury poisoning of movie stars is making headlines is somewhat of a breakthrough for a toxin that has been haunting humans for thousands of years with little profile. In fact, it might just take a Hollywood star to motivate us to get around to protecting the public effectively and taking mercury in our fish and water seriously.

The mercury story is one of human tragedy, industrial malfeasance, government collusion and the shocking inability of humans to act prudently when presented with the facts. There are powerful lessons to be drawn from mercury that help shed light on some of the other substances Rick and I tested. Why, for example, does it take so long for a toxic chemical to be banned from use in products humans consume when we know it causes harm? And how is it that we continue to believe the corporations that profit from toxic pollution when time after time, substance after substance, they are proven to be wrong and the public pays the price? By asking these questions I am hopeful that we are starting to make some progress on the solutions.

Bruce's Tuna Feast

Speaking of losing brain cells, I'm not sure whether deliberately setting out to elevate my mercury levels was a sign of too few brain cells to start with or simply evidence of a healthy scientific curiosity. I'd like to think it was the latter. As a long-time mercury policy advocate, I have a special connection to this experiment. It's one thing to spend ten years telling people mercury is bad news. It's quite another to try to elevate my own mercury levels and actually test whether or not mercury in fish is as great a problem as the scientific studies suggest.

It's Saturday and the tuna eating experiment is about to begin. I have to start by saying I love tuna. I love tuna sandwiches, I love tuna sushi, I love grilled tuna steaks. Because of my love of this fish, and frankly all seafood, I tend to eat more of it than the average North

American or European; my diet is perhaps somewhat Japanese. It's not unusual for me to eat about eight fish meals in a week. I've been known to have about four or five dinners that include mussels, crab, shrimp or scallops—with perhaps a smoked fish appetizer one night. For lunch I might have a tuna sandwich and a salmon sandwich in the middle of the week, or maybe sushi or shrimp pad thai, and one morning a week I might have smoked salmon for breakfast. Fish, after all, is an important source of protein and fatty acid. The message here is certainly not to avoid fish, but to be careful about which fish you eat. Most of the fish I eat is low in mercury. So with the task of figuring out who was going to expose themselves to which toxic chemicals, I, of course, jumped at the chance to be the tuna guinea pig. How difficult could it be to eat a little tuna? And surely eating a few tuna meals over the course of 48 hours is hardly going to affect *my* mercury levels, I thought to myself.

To be clear, given my high level of fish consumption, the first thing I needed to do before testing was to try to bring my levels down to background levels more like those of a typical North American. We did no pretesting at this stage, but to be safe I avoided tuna and most other sources of fish for about six weeks prior to our formal testing. Given the half-life of mercury in our bodies, this probably had a modest impact on my actual mercury levels.

Colleagues of mine at the University of Quebec in Montreal are among the leading mercury researchers in the world. Whenever they do field research, they test their mercury levels in advance of their travels and upon their return. Invariably, their mercury levels increase noticeably as a result of eating the local fish at the places they visit.

I therefore knew that it was theoretically possible to measure the mercury increasing in my blood by eating fish that presumably contained mercury. What I did not know was whether eating a few meals over the course of 48 hours would show any measurable increase in my blood. I was worried that given the large amounts of fish I eat regularly, a few extra tuna meals would not have much effect.

After my first blood test to determine my pre-tuna mercury levels, I got to the task of gobbling down a tuna sandwich. We purchased many varieties of canned tuna, but I chose my favourite, solid white tuna. Solid white also happens to have the highest levels of mercury. Flaked or chunk light tuna has lower amounts of mercury because the fish used in flaked tuna tend to be smaller. Smaller fish have lower concentrations because they are younger and eat smaller fish themselves, so the effects of "biomagnification" tend to be less pronounced. ("Biomagnification" is the term used to describe how levels of toxins increase, the larger and higher in the food chain a predator is—because it keeps not only its own toxins, but also the accumulated toxins of the prey it eats, the prey its prey eats and so on.) The larger fish are also usually older and have had more time to bioaccumulate mercury in their diet over many years. Tuna can live for 20 years and reach weights of up to 1,500 pounds. These are the most prized for sushi.

Most readers will be familiar with the classic tuna sandwich or tuna salad, but we all have our favourite variations. My tuna sandwiches are usually made with a can of tuna, a tablespoon or two of real mayonnaise, a chopped celery stalk and a big squeeze of lemon. Unfortunately, on Day 1 we were out of celery, so the tuna salad was a little bland. I spread the filling on commercial whole wheat bread.

Without celery as filler the tuna seemed to disappear easily into the bread, and before I knew it I had managed to get an entire 7.5-ounce can of tuna into my sandwich. Six minutes later I'd downed it. Rick stared at me with an evil glint in his eyes and then turned to the other tuna cans on the counter.

"Surely," he remarked, "that little tuna sandwich didn't fill you up. How about another?"

The steps above were repeated, and I managed to put back a second sandwich with another entire 7.5-ounce can of tuna in it. Now I do appreciate that this may not have imitated the diet of an average person. But at the same time, it's not completely out of the ordinary. The second one took more than 6 minutes to eat, perhaps 15 or so. That's the only tuna I ate on Day 1.

It was Day 2, another 24 hours had passed before I ate more tuna, and once again I had a tuna sandwich for lunch (and a blood test at 5:15 p.m. to measure my mercury). Rick and I were in the middle of breathing perfluorinated stain-resistant chemicals at the time (see Chapter 3), so the chance to get out to the kitchen was a welcome break from the nasty fumes. This time I ate only one sandwich filled with an entire 7.5-ounce can of solid white albacore tuna. It took a little longer to eat than the previous day's, but I have a hunch it was because I was prolonging my time in the relatively poison-free atmosphere of the kitchen. I also had a cup of tea, but not in polycarbonate plastic (see Chapter 8); deliberate exposure to two or three nasty chemicals at a time is enough for me. Tea, according to a recent paper by my mercury-studying colleagues in Montreal, can remobilize mercury stored in your body.[3] That means it adds old mercury probably from my liver to the new tuna mercury, which seemed to be a useful thing, given the short-term nature of our little experiment.

After a few hours of sitting and breathing PFOA, Sarah, our intrepid project coordinator, decided to pop out and get some tuna sushi take-out. At 5:45 p.m. I downed a healthy trayful of tuna sushi and sashimi. For non-aficionados of this delicacy, sushi is raw fish on a little bed of cold sticky rice, and sashimi is just a piece of raw fish on its own. I ate them with wasabi (Japanese horseradish) and soy sauce. Unbeknownst to me this was Sarah's idea of an appetizer. Soon after happily gobbling down the contents of my tuna sushi tray, Sarah presented me with another tray of (much nicer looking, I must admit) tuna sashimi, sushi roll and nigiri sushi—for dinner. It took me a good 40 minutes to polish off this batch, and I was forced to consume a beer or two with it. Though I love sushi, I must admit that eating a large quantity of the dish in one sitting was a little tough.

Day 3 looked remarkably like Day 2, with another tuna sandwich for lunch. It was not a very memorable sandwich, to be honest, or perhaps that was the mercury kicking in, since I don't remember much of Day 3; frankly, it was a hellish day. I am generally very relaxed, easygoing and almost unflappable, but by Day 3 in our apartment, I was miserable. It suddenly occurred to me that perhaps this feeling of intense and uncontrollable irritability was the mercury building up in my body. I was also in no mood for casual conversation. Was it just a coincidence that irritability and shyness are early indications of mercury poisoning? Was I exhibiting signs of elevated mercury in my blood or was it basic hypochondria? Was I experiencing a Daphne Zuniga "brain fog"? At this point I had no idea whether or not my mercury levels were, in fact, any higher, but I was definitely experiencing an unpleasant anxiety.

Amazingly, despite the perceived health woes, I was looking forward to another tuna dinner. Sarah went out and picked up a couple of big, thick tuna steaks from the fish market. We decided we'd all have seared tuna steaks together, and I was more than happy to cook them. By now—and after reading Chapter 3—you may have guessed that they would be cooked in a frying pan with a non-stick coating. This was not to keep the tuna from sticking, of course, but to make sure we were not missing any opportunities to add to the perfluorinated chemicals Rick and I were inhaling in the other room.

I rubbed the tuna steaks in a combination of white and black sesame seeds and seared them in the hot frying pan with a little olive oil. We had a small salad and a little wasabi mayonnaise and, of course, an icy beer or two. Despite this being my seventh tuna meal or snack in three days, I thought dinner was absolutely delicious. I had no trouble consuming a hunk of tuna that weighed in at just over a pound (500 grams), far larger than a typical serving. Actually, fresh tuna is so expensive that only rich people can poison themselves in this way. At close to $20 a pound, it is definitely not a poor man's toxic dinner.

There are, of course, cheaper ways of elevating your mercury blood levels, especially if you enjoy sport fishing. Most lakes in

North America have fish advisories warning against eating certain fish, and 80 per cent of fish advisories are due to mercury contamination.[4] If you are an excellent angler and can catch good-sized fish, like pickerel or walleye, on a regular basis, you'll have no trouble poisoning yourself. In fact, this happened to a fellow living in Minnesota, who became seriously ill, to the point where he was hospitalized and unable to walk.[5] After numerous tests and medical consultations, one doctor finally asked the man's wife if she could think of anything unusual about what he was doing or eating. She thought for a moment and mentioned that he loved to fish and that he ate much of the fish he caught. In fact, she said, he ate fish virtually every day of the week. The doctor immediately tested his blood for mercury, and sure enough, he had levels high enough to qualify him as having Minamata disease, so named after a town in Japan where mercury poisoning was rampant. More on that infamous incident later in this chapter.

Although I suspected my mercury levels would have increased, only the blood tests would determine whether, in fact, there was enough of the chemical in my seven meals and snacks to make a noticeable difference in the concentration of mercury in my own blood. Usually, waiting for blood test results leads to anxious anticipation or perhaps grave concern. But I was in a state of intense curiosity as to whether my blood levels would have above-normal levels of mercury. As with all of our blood samples, these were taken according to standard medical research protocols and the blood was centrifuged on site. My mercury blood samples were sent to Brooks Rand Labs in Seattle and tested using EPA Method 1631 protocols.

I was familiar enough with the literature to know that it was possible to elevate mercury levels in blood, and I certainly knew my colleagues in Montreal had demonstrated this. But I was still not convinced that seven meals and snacks would do the trick. What would a substantial increase look like anyway? If I increased my mercury levels by 10 per cent, was that really a big deal? What about 50 per cent? Or imagine if my levels doubled! Now that would be impressive, but it seemed impossible, given my ongoing higher-than-average fish consumption before doing the experiment. It's also important to keep in mind that with mercury, there is no safe level. Medical researchers have determined that any amount greater than zero increases the risk of harm to humans.[6]

After five weeks of waiting, my results arrived. My first blood test (taken on Day 1 before I ate my two large tuna sandwiches for lunch), showed that the mercury concentration in my blood was 3.53 µg/L (micrograms per litre). The North American average is less than 1 µg/L, so I had about four times the average mercury levels in my blood before my tuna eating even began. And this was after not eating fish for six weeks. I figured my levels would be a little higher than average but not quite that high.

After three mercury meals/snacks, the mercury in my blood shot up to 7.55 µg/L, more than doubling in less than 48 hours. This also sent my mercury levels above the U.S. Environmental Protection Agency reference dose level of 5.8 µg/L. The reference dose is the "safe" level set by the U.S. government, and anything above 5.8 µg/L is definitely cause for concern, primarily for women of childbearing age.[7] This number is based on extensive research in populations that consume large amounts of fish and marine mammals, such as the people of the Faroe Islands in the North Atlantic.[8]

In my final blood test, the morning after my tuna steak dinner on Day 3, my mercury blood levels reached 8.63 µg/L. After seven meals/snacks in three days, I had managed to *more than double* the mercury levels in my blood! Almost two and a half times, in fact. The experiment had worked. Not only did it reveal high levels of mercury in several sources of tuna, but it also demonstrated how easy it is to spike mercury levels by eating a few sandwiches and a couple of tuna dinners. After reading these results, I got a first-hand understanding of how communities that depend on fish in their diets can quietly poison themselves.

Figure 4. Bruce's mercury blood levels (in μg/L) increase significantly as a result of eating fish.

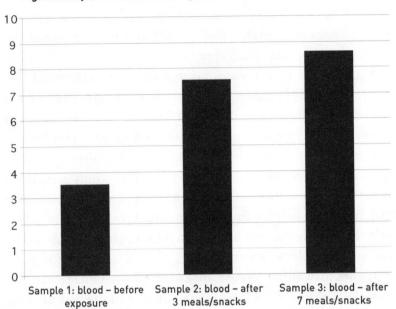

Sample 1: blood – before exposure Sample 2: blood – after 3 meals/snacks Sample 3: blood – after 7 meals/snacks

Mercury and Me

Mercury is cool stuff. For over a decade I've been researching its uses, its effects on health and sources of mercury pollution, as well as supporting government efforts to reduce mercury levels in the environment.

I started working on mercury issues with a Canadian environmental group called Pollution Probe about 15 years ago. At the time most of my environmental colleagues were focused on a major global effort to ban chlorinated compounds. The thinking was that lots of nasty chemicals are based on chlorine. So rather than trying to ban deadly chemicals one at a time, a prospect that could take centuries, the idea was to look at "classes" of chemicals—like chlorinated substances—and seek restrictions on all chemicals in the group that shared certain fundamental toxic properties. In retrospect, it's apparent that this was an intellectually and scientifically sound concept, but at the time politically impossible.

After three years we were making virtually no progress on addressing the problem of complex chlorinated chemicals. The health data were controversial, industry opposition was vigorous and the regulatory systems of Canada and the U.S. could not handle our groundbreaking approach.

This is when it occurred to us that we were trying to take on dozens of complex chemicals for which health effects research was scarce or contentious, while mercury, one of the oldest and most studied toxic materials, was still being used freely in hundreds of consumer products. How is it, we asked ourselves, that given all we know about mercury, we are still putting it in skinny glass tubes and sticking those tubes in people's mouths—and other places, for that matter? Surely, we thought, if we can't restrict mercury use, we're hopelessly doomed in our efforts to rein in any of the newer and more complex synthetic toxins.

I worked with colleagues in Environment Canada and the Ontario Ministry of the Environment to investigate and advocate the restriction of mercury use in Canada. Three of us virtually carried the Canadian mercury file for most of this period. In fact, I recall one day when Ian, my provincial government counterpart, said that if it were not for each of us playing our respective roles— me pushing for policy action and Ian and his colleagues trying their hardest from the inside to move their behemoth bureaucracies— nothing would be happening on mercury in Canada.

We started our research by examining all the sources of mercury entering the environment. It turned out that much of the mercury contaminating our water and fish came simply from the use and disposal of everyday consumer products. Over the next ten years, it became my quest to figure out how and why, after thousands of years of clear evidence of harm, we still allow our bodies to be poisoned with mercury.

Mercury is one of the oldest poisons known to humans. It is one of the most toxic substances to which humans are regularly exposed, it has thousands of uses and it is also one of the best-studied toxins on the face of the earth. For these reasons mercury can help us understand pollution issues more broadly—particularly how it is that something so deadly is still used widely to this day.

In Chapter 1 we described how toxic pollution has gone, more or less, through the following phases. The death by direct exposure phase, the belching and spewing industrial phase, the workplace exposure phase and finally the more subtle poisoning of the entire population with tiny, invisible amounts of toxins in our food and water. Unlike most toxins we tested, mercury has gone through all these stages during its long history.

Magical Mercury Tour

Anyone who has broken a thermometer knows how strange and fascinating it is to watch the mercury inside turn into tiny, silver-coloured balls that split apart and scatter. Even more bizarre is pushing the tiny balls back together, taking ten or so separate little blobs and watching them recombine into one perfect, larger sphere. It's an amateur alchemist's dream, the human equivalent of crows being drawn to shiny objects.

Given these alluring attributes, it's no wonder that many cultures believed mercury had mystical properties, including the power to prolong life. In certain Latin cultures it is used to this day in attempts to ward off evil spirits. Mercury amulets can still be purchased at street markets throughout Mexico and parts of Central America. Perhaps one of the more dangerous practices was sprinkling liquid mercury in infants' beds to protect them from evil spirits and keep them healthy. This practice was also carried out in Latin communities in New York City until very recently, causing public health workers to initiate special education programs warning parents of the serious danger this poses.[9]

Mercury has been used for many other purposes over the centuries.[10] A Danish researcher has recently discovered that six medieval monks appear to have died of mercury poisoning from using mercury-based inks to transcribe religious documents.[11]

The monks probably licked their brushes to make the beautiful fine lines with mercury-laden red pigments that are still so vibrant in their illuminated manuscripts.

The Ancient Romans discovered that mercury combines with gold and other precious metals and took advantage of this property to recover gold and silver in the early days of mining. During the Renaissance the physicians of the day used its healing properties. Throughout the American Civil War, mercury was considered to be a "cure-all" for everything from skin lesions to constipation. Abraham Lincoln was prescribed mercury tablets, called "calomel," but being the smart fellow he was, he soon recognized the telltale signs of mercury poisoning and stopped taking them.[12]

Research was conducted on Spanish mercury miners in the 1960s that included memorable images of the miners' attempts to trace a curved line on paper.[13] The miners were unable to follow the curve at all and instead produced a jagged, squiggly line resembling a child's attempt to draw lightning. Uncontrollable trembling is one of the early signs of serious mercury poisoning.

We are all familiar with stories of Spanish galleons carrying tonnes of gold and silver looted from the Native people of Central and South America. Much less well known is the fact that many tonnes of mercury were transported in those same ships from Spain to the "New World." After murdering many of the locals and melting the golden riches of the Aztec and Mayan cultures into gold bars and coins, the Spanish conquistadors used mercury to mine even more gold, and silver. Thousands of tonnes of mercury were shipped from Spain during the four-hundred-year period of Spanish rule. My colleague, Luke Trip from Environment Canada, visited Zacatecas, Mexico, where up to thirty-four thousand tonnes of discarded Spanish mercury may be present.[14] Luke said it's possible to scoop a handful of dirt and squeeze liquid mercury from it. To put this in context, one tonne equals 1 million grams, and if the conditions are right, one gram can contaminate the fish in a 20-acre lake. It's almost impossible to imagine the suffering associated with the billions of grams of mercury the Spanish brought with them to Latin America.

The height of 18th-century European fashion was the beaver felt hat, or what we think of as a typical black top hat. Owing to its fungicidal properties, mercury was used in the manufacture of these hats to ensure that the fur would not "go off." Unfortunately, the workers in the beaver felt factories of the day did not fare quite so well. In fact, the hatters would often go mad as a result of breathing the toxic mercury fumes—hence the phrase "mad as a hatter." The fact that irritability is one of the early symptoms of mercury poisoning may also account for the "mad" moniker. The most famous "Hatter" is most certainly the crazy, riddle-filled character in Lewis Carroll's *Alice in Wonderland*.

Mercury pollution these days is more complex. We now worry less about direct exposure causing disease and potential death and more about long-term exposure to tiny amounts in our food that can cause insidious neurodevelopmental problems, especially in children.

Mercury is different from the other substances Rick and I are writing about, because it is a naturally occurring element, not a manufactured chemical. It's been around forever—literally. Elements are the building blocks of natural and human-engineered chemistry. All the other substances that Rick and I tested were synthetic chemicals—products typically created in laboratories by chemists working for chemical companies. Mercury is very different in that it cannot be created—or destroyed, for that matter. It simply exists as a natural element found in rocks, plants, water and most living things. So volcanoes, forest fires and oceans all release mercury into the atmosphere. This is called "natural" mercury, because the source of the release into the environment is natural. There is also "anthropogenic" mercury, which results from human activity. All mercury generated by human-made sources (such as waste incinerators, coal-fired power plants, mercury thermometers and fluorescent lights) is called anthropogenic. Where it comes from may not seem to matter all that much, but it actually does, and I'll explain why a bit later.

Is Mercury Really Dangerous?

Whenever I mention mercury in casual conversation, which happens quite frequently, the first thing people often ask is "I used to play with mercury as a kid whenever a thermometer broke. Was that dangerous?"

"Well, probably not really, as long as you didn't play with it for hours on end, day after day," I'd reply, knowing that lots of people, like me, did.

Then I tell the story of the dentist in British Columbia who decided to heat mercury on his stove.[15] It seems he had a great fascination with mercury, and for some unknown reason decided to boil some up. Mercury vapour is highly toxic and the vapours killed him. Not only that, his apartment building was condemned due to the mercury levels, forcing all the residents into a local motel.

Then there was the boy from Ohio who was hospitalized with a mysterious, debilitating illness, later discovered to be mercury poisoning.[16] It turns out there had been an accidental mercury spill in his family's apartment, and the mercury vapours caused the 15-year-old to become very ill. His symptoms included a rash, sweating, cold intolerance, tremors, irritability, insomnia and anorexia. He was diagnosed with measles, sent for psychiatric treatment and even accused of having psychosomatic symptoms before mercury poisoning was identified as the cause.

In another case a nine-year-old boy was being treated for neurological and kidney problems after a blood pressure device broke, spilling mercury in his house.[17] Unfortunately, the boy's mother vacuumed up the spilled mercury, thus inadvertently turning the vacuum cleaner into a mercury vapour distribution device. Each time the vacuum was operated, the warm air from the vacuum sprayed highly toxic mercury vapour throughout the apartment.

And consider the terrible tale of a chemistry professor at

Dartmouth College.[18] Karen Wetterhahn was very concerned about the toxic effects of mercury on humans and the ecosystem and was working with a particularly toxic form of the metal called dimethylmercury. This is not a common form, but it is used in research because it produces the effects of mercury very rapidly, allowing studies to take place over short periods of time. Special precautions are required when handling it, and Wetterhahn was wearing latex gloves and working under a fume hood to protect herself. It is believed that one or more drops of the lethal liquid spilled onto her latex gloves. The mercury leaked through the latex and reached her skin. The tiny amount of mercury was quickly absorbed by her body and led to devastatingly high mercury levels in her blood. Wetterhahn knew the effects of mercury poisoning and witnessed firsthand some of the classic symptoms: shaking, slurred speech and tunnel vision.

Less than a year after this exposure, Wetterhahn was dead. Her brief exposure to mercury represented nearly one hundred times the lethal dose. This young woman, who had devoted her life to preventing mercury from harming humans and the environment, died as a direct result of her concerns. The most tragic part of the story is that if governments and industry had acted responsibly over the past 50 years and banned mercury uses when it was obvious they were causing harm, she probably would not have needed to be doing this research in 1996.

Exposure to high levels of everyday mercury (let alone dimethylmercury) has serious consequences. It can cause permanent brain damage, central nervous system disorders, memory loss, heart disease, kidney failure, liver damage, cancer, loss of vision, loss of sensation and tremors. It is also among the suspected "endocrine disruptors," which do damage to the reproductive and hormonal development of fetuses and infants. Some studies also suggest that mercury may be linked to neurological diseases, such as multiple sclerosis, attention deficit disorder and Parkinson's, but the evidence here remains somewhat inconclusive.[19]

Medical researchers at the University of Calgary have identified the exact cause of mercury damage to the brain.[20] Mercury actually concentrates in major organs like the brain and kidneys, dissolving the neurons in certain parts of the brain and leading to various nervous system disorders. The Calgary researchers created a video (now on YouTube) that shows the mercury molecules literally destroying brain neurons as though they were little Pac-Mans munching away on brain cells. Alzheimer's and autism are associated with brain neuron damage.

So in case there is any doubt, mercury is a seriously potent neurotoxin that will kill you if you breathe it, eat it or otherwise expose yourself to high enough levels. Even moderately high levels taken in over an extended period of time will cause serious physical and mental impairment.

Cool Liquid Metal

Mercury has an unbelievable array of unique and fascinating properties. Perhaps its coolest property is that it is the only metal that is a liquid at room temperature. Think about that for a minute: liquid metal. Most other metals need to be heated to hundreds or thousands of degrees Celsius before they melt. Most of our images of liquid metal are of refinery workers in protective suits and helmets holding long tongs to pour blazing liquid metal into ingot casts. Liquid mercury, on the other hand, can be stored safely in a plastic bottle.

As with other metals, mercury conducts electricity, but the unusual combination of its being liquid *and* an electrical conductor has led to its use in electrical switches. Remember the round household thermostats that first came into use in the 1950s? Or how about the silent wall switches popular in the 1970s? Just one of either of these devices contains enough mercury to contaminate a 20-acre lake to the point where the fish in the lake should not be eaten.

So why is there mercury in switches? It's quite simple: The little blob of liquid mercury slides back and forth each time the switch is flicked, resulting in an electrical contact forming or being cut off and the light turning on or off. Check to see if you still have a silent light switch at home. It's easy to tell because rather than clicking on or off, they're very smooth and quiet, and they give almost no resistance to being flicked. They can be quite easily and safely taken apart to reveal one or two silver metal disks about the size of a dime, which contain about one gram of mercury each. When you pop out the disks, you can shake them and feel the weight of the mercury sloshing around inside. If you do find mercury switches in your home, make sure they are returned to a hazardous waste depot, not discarded in the garbage.

You might think that mercury switches are a pretty limited application until you realize that so-called "tilt switches" are everywhere. Anyone for ice cream? When you open up the freezer lid, a little blob of mercury slides down the switch and, presto, the light comes on. How about trying to find the tire iron in the trunk of your car? Luckily, a little light goes on when you open the trunk, all thanks to tilt switches and the mercury they contain.

If you are (un)lucky enough still to have a classic round thermostat on your wall, you can actually watch this firsthand. Simply grab the sides of the thermostat and gently pull off the curved, circular cover, which is usually gold-coloured plastic with a large hole in the centre where the temperature disk is. It looks something like a doughnut. Once the cover is off, you will see a glass tube under some wires at the top of the thermostat. When you turn the temperature gauge up or down, you'll see the glass tube tilt back and forth and you'll also see the shiny liquid mercury sloshing from one side of the tube to the other. When you turn the heat up, the switch tilts, the mercury slides down and makes electrical contact, and the furnace turns on. This is a tilt switch in action.

Mercury has other uses as well. Because it's a highly volatile substance and evaporates to form vapour easily, it is used in virtually all fluorescent and neon lights. The metal vapour conducts electricity inside the glass tube, causing various gases to fluoresce when an electrical current is introduced. Mercury also has the unique property of expanding evenly with pressure and temperature, and this makes it perfect for thermometers, barometers and manometers.

Mercury can be used with other metals, creating combinations called amalgams. (The other metals dissolve in mercury like salt in water.) The best-known mercury amalgams are made with silver and are used to fill cavities in teeth. The average "silver" mercury filling is one-half mercury by weight. Some dentists continue to use mercury, but it is largely being replaced with white composite resins.[21] Controversy and active debate surround the health effects of mercury fillings. A segment of the population appears to have hypersensitivity to mercury, and these people often have their mercury fillings replaced (although removing a large number of mercury fillings at one time is not advisable, since that can result in very high temporary mercury exposure).

Sticking mercury in our mouths really does strike me as one of the crazier things we decided to do with it. I spoke with Dr. Peter Erickson, a Canadian physician currently living in Texas, who specializes in treating patients with environmental sensitivities, including mercury, and he told me the story of how it became the most popular dental filling material in North America. Mercury's toxicity was well known at the time and it was considered too hazardous for dental fillings. But it was easy to use, much cheaper than gold and had the added benefit that its toxicity killed any bacterial infections.

In 1833 the French dentist who had perfected mercury amalgam use was banned from practising in France, so he moved to New York, where he opened a highly successful dental practice.[22] In the mid-1800s dentists in the United States belonged to the American Society of Dental Surgeons. The society, following European standards of the day, banned mercury use by dentists. According to Dr. Erickson, the French dentist and his brother, together with a number of other mercury-using colleagues, decided that the only way they could continue to use mercury was to leave the Society of Dental Surgeons and create their own new organization, called the American Dental Association. The American Dental Association (and its Canadian counterpart, the Canadian Dental Association) became the *de facto* proponents of this practice in dentistry, ignoring the health concerns first expressed over 150 years ago. To this day the American and Canadian Dental Associations remain among the staunchest defenders of mercury.[23] They have vigorously opposed any efforts to restrict mercury use, and they have even fought proposals to make information about mercury available to patients. But despite the opposition and foot dragging, the tide has now turned, and the use of mercury amalgam is finally decreasing.[24]

Perhaps mercury's most obvious property is that it kills living things, including mould and fungus. These properties were well known in the 15th and 16th centuries, and ailments ranging from constipation to ringworm fungus to syphilis were treated with mercury into the early 20th century. This has led to speculation that some of the crazy antics of famous emperors and leaders of the past may have been caused by mercury poisoning from their syphilis treatments.

Throughout the 20th century mercury was used widely in bathroom, kitchen and hospital paints to prevent the growth of mould and mildew, and mercury emissions from drying paint were significant. Most Western countries have now banned this practice. In countries like Canada, where the regulation of toxic substances has been virtually unknown until recently, manufacturers follow international standards. (The Canadian market is relatively small—compared to markets in the U.S., Europe or Japan—so Canada rarely sets industry standards that exceed those of other nations. We have not technically banned mercury from paint, but companies operating in Canada agreed voluntarily not to sell paint containing mercury.) Mercury has also been used as an agricultural fungicide, notably on potatoes. The high frequency of unusual cancers among potato farmers in Prince Edward Island may be linked to specialized potato fungicides and pesticides.[25]

Given the well-known toxicity of mercury, it is surprising that it still has so many applications. We put it in children's vaccines and nasal spray and even in contact-lens solutions. Happily, in the past few years, most of these uses have either ceased or become severely restricted. The only one that still remains is—you guessed it— injecting it into babies, in the form of vaccines. Fortunately, however, this practice has come under examination of late.[26]

Dancing Cats and Human Tragedy

The devastating effects of mercury poisoning burst onto the international stage in 1956 when residents of Minamata, Japan, started to become very, very ill. Strange behaviour in cats was the first sign that something was awry in this fishing town on Japan's southernmost island, Kyushu. Throughout the town cats were literally jumping, twisting and doing back flips, which led to the term "dancing cat disease" in reference to the uncontrollable muscle spasms and tremors seen in the poisoned felines.[27] Further research on the cats led local health scientists to the conclusion that mercury contamination in fish and shellfish was the cause of this strange and lethal disease. In addition to dancing cats, seabirds were dropping from the sky, unable to fly.

Seafood was, and still is, the primary protein in Japan and the most important part of the diet of any fishing village. Along with the cats, people in Minamata started to show the telltale signs of mercury poisoning, including trembling, numbness, irritability and tunnel vision, but at the time mercury poisoning was not commonly known in the medical community.

The Minamata poisoning episode provided local medical researchers with the hard evidence that first linked mercury pollution in water to mercury contamination in fish and ultimately mercury poisoning in humans.[28] The tragedy in Japan happened ten years before the modern environmental movement began and in the very early days of even the most rudimentary understanding of ecosystems. But local medical specialists were still reasonably quick to identify the cause of the poisoning as methylmercury, based on studies of the deceased, no-longer-dancing cats. It is important to point out that methylmercury is the organic form of mercury, making it much more dangerous in food. This is because organic chemicals ("organic" used here in the sense of substances composed of carbon and hydrogen atoms, not food produced on environmentally friendly farms) can be most easily incorporated into human blood and tissue.

Once the methylmercury link was made, it was not long before the source was discovered to be a chemical plant in Minamata that manufactured polyvinyl plastic. The mercury-laced industrial waste was being dumped directly into Minamata Bay. The same bay where local fishers placed their nets and traps.

Tragically for people in Japan's fishing industry, methylmercury bioaccumulates and biomagnifies more powerfully than almost any other substance known. Even at very low rates of bioaccumulation or with relatively low concentrations of mercury in the water, biomagnification can result in toxic mercury levels in fish. Top predator fish can have mercury concentrations that are hundreds or thousands of times, possibly even a million times, greater than concentrations in the water in which they swim. This is why large ocean predators such as shark, tuna, swordfish and marlin have the highest mercury levels. And this was why eating large fish out of Minamata Bay was deadly.

Eating poison fish, sadly enough, is not the greatest tragedy at Minamata. The most despicable part of this episode was the response of the Japanese government and of the chemical company responsible for the mercury dumping. Government officials ignored the well-established evidence of the local medical researchers for more than 10 years. Without any government action and absent the concept of corporate responsibility, the chemical company continued to poison the people of Minamata for all those additional years. The Japanese government refused to acknowledge a connection between the mercury poisoning and the deaths and suffering of thousands of Japanese citizens, even when the medical evidence was clear. Stillbirths, serious deformities and the poisoning of tens of thousands of people may have been largely avoided had the government and the chemical company not acted with such blatant disregard for human suffering.

Today, the symptoms of severe mercury poisoning are still referred to as "Minamata disease." In Japan court battles continue even now between the citizens of Minamata and the Japanese government over compensation for the poisoned families. Despite the devastation and gross negligence on the part of the government and the chemical company, the Japanese government is still fighting to minimize the official estimate of those affected, thereby limiting the compensation it might pay. The official government line is that 2,265 people were poisoned by methylmercury. Kumamoto University researchers put the number at 35,000. Many severely debilitated survivors are still living in Minamata today, but many, many more are no longer with us.[29]

Paper, Rock, Fish

Soon after the tragic mercury poisonings in Japan, a number of serious incidents occurred in North America. In 1969 a pulp and paper mill polluted the English-Wabigoon River in northern Ontario, contaminating the water so severely that the fish were no longer safe to eat. Not only did this destroy the primary food source for the local people; it destroyed their traditional way of life.

Mercury is used in what are called chlor-alkali plants as part of the pulp and paper manufacturing process. It was common practice to have a mercury cell chlor-alkali plant connected to a paper mill. It was also common practice for tonnes of mercury to be dumped into local rivers from these plants.

Testing showed that the White Dog and Grassy Narrows First Nations people exhibited high levels of mercury in their blood and hair, but there is some dispute as to whether any of them exhibited levels high enough to suggest symptoms of Minamata disease.[30] Federal and provincial governments claim the mercury levels were only modestly elevated, since the people were warned in time not to eat the fish. Some independent studies suggest otherwise, although there is no conclusive evidence. The bigger issue is, again, the negligence that led to enormous quantities of mercury being dumped into the river system at a time when mercury poisoning in Japan was making international headlines. Similar incidents were occurring across North America, causing the closure of commercial fisheries and destroying the food supply for dozens of local communities.[31]

The mercury from chlor-alkali plants is elemental mercury, meaning it is pure, metallic mercury, as opposed to the much more toxic organic form, methylmercury, which was being dumped directly into the ocean at Minamata. The English-Wabigoon River fish were, however, contaminated with organic methylmercury similar to that found at Minamata. How could this be so? you may ask. The answer requires a final chemistry lesson. To make a long story short, mercury undergoes a process called "methylation," and methylation is critical to understanding how mercury ends up in fish—and cats and birds and humans, for that matter.

Methylation is the process through which mercury is converted from "inorganic" to "organic" mercury. "Organic" mercury contains carbon atoms, making this form of the metal much more absorbable by living things. Methylation occurs naturally and to a significant extent in lakes and rivers around the world if the conditions are right. In fact, most lakes, especially northern lakes in North America, contain methylmercury. Places like Minnesota, Ontario, Quebec and Wisconsin seem to be particularly well suited to methylmercury formation, and this has to do with a number of factors, including the type of rock, the acidity of the water and the presence of organic matter in lakes where mercury is found.

To explain the rest of the process, let's look at the reservoirs of hydroelectric dams, where elevated levels of mercury were found in fish in the late 1970s. This type of pollution still exists in dam reservoirs today and is of special to concern to Canada, where so many hydroelectric dams are in operation. Unlike the mercury from chemical plants in Japan or pulp and paper mills in North America, the mercury in hydropower reservoirs is not dumped directly into the water. It comes from soil. Mercury levels in lakes are also affected by additional mercury that is deposited with the rain, mainly blowing across the North Pole from coal-fired electric plants in China and Eastern Europe.

Mercury is a naturally occurring element found in rock and soil, and it is affected when a river is dammed. First, the dam causes a large area of land to be flooded. Next, when dams flood large areas of forest, the trees die and decompose in the reservoir. This is a critical step, because the rotting plants and animals produce perfect conditions for methylating micro-organisms. The mercury found in the rock underlying a hydro reservoir is released by methylation caused by the increased bacterial activity associated with the decomposition of plant life in flooded areas. And with increased methylation come elevated levels of methylmercury. In northern Quebec, where some of the world's largest hydro dams are found, the mercury levels in large predatory fish in the dam reservoirs are far too high to allow the fish to be consumed by the local Cree population for several decades after contruction.[32]

As mentioned before, methylmercury is the form of mercury that most easily enters our bodies. From a health risk perspective, a toxic substance is only as dangerous as its ability to get inside your body and harm critical bodily organs and functions. Methylmercury has these characteristics in spades, including two of the most serious toxicity traits: It crosses the blood-brain barrier and it crosses the placental barrier. So despite our bodies' best efforts to keep nasty things from getting into our brains and our unborn babies, methylmercury slips through with ease. In fact, not only is it able to get into our brains, but mercury seems to prefer hanging out in our grey matter—hence the term "neurotoxic." Mercury also binds to proteins. (This is different from "lipophilic" chlorinated chemicals such as pesticides or PCBs that are stored in our fat.) So in addition to collecting in brains, mercury tends to concentrate in major organs such as the heart, liver and kidney, and kidney failure is one of the major causes of death from mercury poisoning.

In case it is not now obvious, the mercury in the English-Wabigoon River was converted to methylmercury by the micro-organisms in the river. Then the fish (the ones that survived, that is) soon became too contaminated to be eaten, and the locals lost their food supply and a large part of their livelihood. Meanwhile, we got nice white paper.

Getting Polluted Is Easy

Even when not eating massive amounts of tuna for purposes of experimentation, I participate directly in mercury pollution; we all do. Some more than others.

The Inuit people of the Canadian Arctic live in what is considered to be the most pristine and fragile ecosystem on earth. Sadly, it is also the world's toxic tailpipe. Poisonous chemicals of all kinds, including mercury, end up concentrating in the Arctic because of global weather patterns and the nasty emission sources located in the northern hemisphere. Arctic animals such as whales, polar bears and seals also happen to be large, long-lived fish-eaters, making them prime candidates for high levels of mercury. In some Arctic communities in Canada, nearly one-third of the women have mercury blood measurements higher than the levels the World Health Organization deems to be a concern.[33] This is in addition to their toxic levels of PCBs, dioxins and fluorinated chemicals. Governments have largely abandoned the Inuit cause, and the alternative of eating frozen chicken dinners and other substitutes instead of local wildlife is neither appealing nor healthy.

The major sources of mercury pollution today are atmospheric, and the two largest atmospheric sources are coal burning, to produce electricity, and waste incineration. In the case of incineration, most of the mercury comes from the products that are discarded, including fluorescent lights, old batteries, drywall with mercury paint and electronics. The mercury in coal occurs naturally, and it is released to the atmosphere up the smoke-stacks of coal-fired electrical plants when the coal is burned. Once in the atmosphere mercury can travel thousands of kilo-metres and can be deposited far from the original source.

The atmosphere is not a great place to practise the "dilution is the solution to pollution" motto (which was popular in the 1980s). It's that kind of thinking that has led to the mercury pollution problems we face today. Once in the atmosphere mercury can also circle the globe, depositing itself in rain and snow, but areas downwind of major pollution sources are hardest hit. In general, mercury levels increase from west to east across Canada, following the paths of prevailing winds and deposited downwind from coal-fired electrical plants and waste incinerators.[34]

Medical researchers discovered that mercury poses a health risk even at very low, continuous doses at about the same time that scientists studying atmospheric mercury shifted focus from direct local emissions to pervasive global pollution. Teams of medical researchers studied children who live in the Faroe Islands (in the North Sea) and the Seychelles Islands (in the Indian Ocean).[35] They chose these places because they're far away from any direct sources of mercury and because fish is a major part of the local diet in both locales. After years of study they determined that there is no safe level of mercury. The old way of thinking (that there is a "safe threshold" and that we can pollute up to that level with no effect) was finally dismissed. Studies, the most famous of which were carried out by Dr. Philippe Grandjean and his team of researchers in the Faroe Islands, found "cognitive deficit" and "impaired motor skills" in children with very low levels of mercury in their blood.[36] The mercury the children were consum-ing did not come from a local factory dumping waste into or near fishing grounds; it was simply the background mercury found in the ocean today.

These studies led to a major revamping of mercury health standards and the issuing of special bulletins around the world, warning pregnant women to not eat *any* high-mercury fish while pregnant. The studies indicated that fetuses and infants are susceptible to even the tiniest amounts of mercury in their developing brains.[37] The Catch-22 is that for many women around the world, local fish is one of the most important sources of protein and omega-3 fatty acids, so not eating fish during pregnancy may be more harmful to an unborn baby than eating mercury-contaminated fish. Not a great choice. And in spite of the risks, Aboriginal peoples of both genders in Canada have been advised to continue eating contaminated fish because of its importance as a protein source.

So where do all these observations and edicts leave us today? There is good news and bad news. The good news is that after thousands of years of direct experience with one of the most toxic substances known, we are finally starting to act intelligently. My concern 15 years ago was that if we couldn't start restricting mercury use, I wasn't hopeful for the success of any reduction in the use of toxins. But thanks to government regulations (mainly in Europe and the United States), mercury use in consumer products has dropped dramatically over the past two decades. Most batteries, paints and switches are now mercury free. Mercury thermometers and thermostats are being phased out. Even dentists seem to be catching on.

Fluorescent lights still contain mercury, although much less now than they did ten years ago. The main challenge is the recent popularity of compact fluorescent lightbulbs (literally billions will be sold). They save energy (and if your electricity comes from coal, they lower mercury emissions), but they all have small amounts of mercury inside them. Proper recycling programs can recover most of the mercury when the lights are discarded, but to date programs like this are not widespread.

The bad news is that coal burning continues to increase at a fierce pace, particularly in China, but the location of the pollution does not really matter, because mercury is a global pollutant. So despite dramatic decreases in mercury use in many consumer products, global mercury levels continue to rise. Without regulations on mercury emissions from coal plants, we will continue to poison important food sources for vulnerable populations around the world.

Tuna lovers, sushi eaters, pregnant women and children should seriously limit the high-mercury fish they eat. Unfortunately, governments are offering no help in this regard. In Canada, fish with mercury levels that exceed the government's own health guidelines can be purchased at any fish market. My tuna steaks were almost certainly in this category. As it turns out, the fish that are most likely to exceed levels in Canadian federal government health guidelines aren't included in the guidelines.

I asked Canadian federal government officials to explain how this was possible and how they could rationalize exempting tuna. I was told tuna are considered by the Government of Canada to be "exotic specialty fish." The Ferraris of the piscine world, I suppose. The thinking behind the exemption seems to be that fresh tuna is so expensive that the average person cannot afford to eat enough to poison themselves.

I think I proved that theory wrong.

SIX: GERMOPHOBIA

I don't like germs. That's why I don't like to shake hands. You just never know what that person did with his or her hand right before it was offered to you to shake. . . .

One final germ warning. Avoid touching the first floor button on the elevator. It is absolutely swarming with germs. I think from now on, I'm taking the stairs.

—DONALD TRUMP, 2006[1]

THERE'S TRICLOSAN IN my garden hose.

Of all the chemicals we're writing about in this book, triclosan was the only one that I was feeling a bit smug about. If you look hard enough, its presence as the active ingredient in many "antibacterial" products is usually labelled, and my family and I have been shunning it for years. So I was pretty sure I had completely banished it from the house. In this one respect I had things under control (or so I thought).

But there I was one evening, watering the little vegetable garden that I had planted for the first time this year and looking down at the hose. I had never examined it closely (I mean, who closely examines their garden hose?), but I noticed for the first time that there were words written on it. Because the letters were so small and the words were repeated over and over again the entire length of the hose, they blended together into what looked like a solid stripe. But as I brought the hose up to my eyes, wiped

159

off the grime and stared hard, I could just make out the phrase "Microban Protection." Microban is an antibacterial product that most often contains triclosan as the active ingredient.

Unbelievable! I looked up at the tomato plants that I was unknowingly dousing with germicide, courtesy of my decidedly un-green, green rubber hose. As I watched the triclosan water soaking into the soil in my backyard, I felt for a moment a wave of extreme exasperation. Perhaps I'm wrong, but I have a hard time believing that consumers are suffering from a plague of germy garden hoses in need of a laboratory-engineered solution!

That evening at dusk, I'm pretty sure my face was actually flushed.

"For the love of God," I thought, "is there no use for these toxic chemicals that isn't stupid and pointless? Is there no corner of our lives that hasn't been invaded by chemical companies peddling their modern-day snake oil?"

Apparently not . . .

Ciba's Baby

Dr. Stuart Levy, a Professor of Microbiology at Tufts University, chuckled sympathetically when I told him about my hose. He agreed that the craze for "antibacterial" products has got out of hand and triclosan has indeed crept into some ridiculous places. "We have antibacterial chopsticks here in Chinatown. . . . Toyota advertises antibacterial steering wheels and certain other features. You've got a hose. I've seen a hot tub. I mean, come on, already. If you're really going to advertise it as a product for health, then put it in something where it's going to work. Microban has succeeded in putting it in everything. You can now get a total triclosan bedroom, complete with pillows and pillow cases and slippers!"

So successful have the purveyors of triclosan been, in fact, that the list Levy quickly rattled off is just the tip of the iceberg. Like my hose some of the consumer products that now contain tri-closan are downright surprising. The Environmental Working

Group has found the chemical in household items as disparate as liquid hand soap, toothpaste, underwear, towels, mattresses, sponges, shower curtains, phones, flooring, cutting boards, fabric and children's toys. One hundred and forty kinds of consumer products in all.[2]

And this is by no means an exhaustive description. When the Canadian Broadcasting Corporation examined the issue in 2006 and 2007, it found that the federal government had, by that point, registered 1,200 brands of cosmetics containing triclosan. Pretty impressive growth for a product that was first introduced in 1972 for use in surgical scrubs and that spent most of its life confined to specific applications inside hospitals and scientific laboratories.

In many ways the history of triclosan resembles that of the brominated flame retardants I talked about in Chapter 4. They are both products in perpetual search of new (and increasingly ridiculous) applications. For decades brominated compounds were used largely as additives for leaded gasoline. When leaded gasoline fell into disrepute—No Problem!—the companies involved started marketing their bromine for use as agricultural chemicals and fire retardants. Using flammability regulations as their tool, they continue to try to insinuate them into everything imaginable.

The triclosan story is similar. Microban and other companies realized they could take a chemical that had previously been limited to hospital applications, build the term "antibacterial" into a saleable brand, water down the chemical's concentration and insert it into products as diverse as deodorant and countertops.

The slogans that are now used to sell triclosan extol the many virtues of the germ-free life. "Spread love, not germs," said the U.S. Soap and Detergent Association (SDA) in one Valentine's Day press release. An advertisement for pet shampoo says that its "gentle yet effective antibacterial action and the crisp scent of fresh green apples destroys odor and leaves your dog's coat clean and shining." And another company tells us to "wipe-out acne bacteria and excess oil with these towelettes. Medicated with antibacterial Triclosan and Salicylic Acid to help to prevent future breakouts."

"It reminds me of the Listerine story," said Katherine Ashenburg, author of an irreverent history of human hygiene called *The Dirt on Clean: An Unsanitized History.* Invented in 1879 from a concoction of thymol, menthol, methyl salicylate and eucalyptol, Listerine was originally marketed as a surgical antiseptic. But then, in a Microban-style manoeuvre, Listerine's makers decided to redefine their product. "They started advertising it for new purposes without ever changing the recipe, the labelling, the price, the bottle, the anything," Ashenburg explains. In the 1890s it was sold to dentists as an oral antiseptic. In 1914 it was first sold as a mouthwash to the general public. "They still weren't satisfied with their sales," Ashenburg continues, "so in the early 1920s the company President asked the company chemist to write him a list of uses for Listerine. One of the things he came up with was that Listerine could be used against halitosis. Well, the President of the company didn't even know what halitosis was. Nobody in America did. Listerine's advertising had to keep defining the word for the next five years."

But the result of defining this new condition and presenting Listerine as the cure was well worth the trouble: Listerine's sales skyrocketed 200 per cent. "I guess clever chemists can find new uses for things or convince us there's a new need for something that's been around for a long time," concluded Ashenburg.

It was the growing use of triclosan that first rang alarm bells for Stuart Levy. In addition to being the Director of the Center for Adaptation Genetics and Drug Resistance at Tufts University, Levy founded and continues to serve as President of the Alliance for the Prudent Use of Antibiotics (APUA). As explained on its website, APUA's mission is to "strengthen society's defences against infectious disease by promoting appropriate antimicrobial[3] access and use and controlling antimicrobial resistance on a worldwide basis."[4]

This is no easy task. Infectious microbes are wily beasts. And the way they adapt to antibiotics is a constant challenge for doctors. APUA recognizes it has a major job on its hands, given that "antimicrobials are uniquely societal drugs because each individual patient use can propagate resistant organisms affecting entire health facilities, the environment and the community." As a result "wide-scale antimicrobial misuse and related drug resistance is challenging infectious disease treatment and healthcare budgets worldwide."

This was the backdrop against which Levy first started noticing antibacterial products and wondering whether their misuse might also be contributing to resistant bacteria. "I got into it because of my interest in antibiotic treatment, antibiotic use and the Alliance for the Prudent Use of Antibiotics. It was at the very beginning of this [antibacterial] phenomenon and Hasbro was impregnating triclosan into some of their toys and claiming it would protect kids from infectious disease transfer. Then [kitchen equipment retailer] Joyce Chen came out with her impregnated plastic cutting board. . . . Corporate marketers discovered that the use of the term 'antibacterial' would be a good marketing ploy; so they started advertising it. I mean one of the funny parts was when you looked at the Reach toothbrush, the triclosan wasn't in the bristles, it was in the handle. And yet it was advertised as the antibacterial toothbrush."

Like a few of the microbiologists interviewed for this chapter, Dr. Philip Tierno, the Director of Clinical Microbiology and Diagnostic Immunology at New York University's Medical Center, revels in telling germy tales. He's been on *Oprah* and MSNBC's *Today* show talking about germs and where we encounter them in our daily life. Tierno also dates his "getting into" triclosan to the Hasbro toys incident. Unlike Stuart Levy, however, Tierno thought Hasbro's innovation was interesting and potentially useful and was quoted as praising it at the time. Though in favour of triclosan in some applications (he personally uses triclosan toothpaste and soap), Tierno is withering in his criticism of others. "There are certain products that incorporate triclosan because they want to jump on the microbial bandwagon and make money rather than preventing transmission of infection from one person to another," he says. "One in particular is a pizza cutter which has a wheel—a metal wheel that you would use to slice pizza—and a plastic handle, and the plastic handle has the triclosan incorporated into it." Tierno calls this an "example of a useless product," given that you have to wash the cheese and other pizza bits off the wheel anyway and therefore by definition have to wash the handle as well.

Tierno adds that "many of the formulations of triclosan contain either too little triclosan to be effective or contain it in a ratio that is not ideal for maintaining its germ-killing ability." He'd like to eliminate these products, he says, "because they are taxing the environment—both the human environment and the environment at large—with unnecessary extra levels of triclosan over and above that which is useful from an antibacterial standpoint."

Interestingly, even the company that invented triclosan has become queasy at some of the uses to which its chemical is being put. Klaus Nussbaum is the global head of the Hygiene Division at Ciba Inc., the company that first introduced triclosan for use in hospital surgical scrubs in 1972. Ciba remains the dominant manufacturer of the chemical worldwide. I spoke with Nussbaum at length over a crackly speakerphone (one of the company's PR people was listening in on the interview) from his office in Basel, Switzerland. Not surprisingly, he was quite positive about the chemical. He pointed out that it's been in use for 40 years and rhymed off a number of papers that have pronounced it safe for widespread use. Near the end of the interview, however, Nussbaum made a passing reference to triclosan disappearing "from some applications we're not in favour of." When I pressed him on this point and expressed surprise that there were any uses of triclosan Ciba wouldn't support, given the company's oft-expressed view that the chemical is not harmful to the environment or human health, he said that it was a "positioning issue for the product." He singled out "widespread, one-use" items like triclosan-infused garbage bags as being of concern to the company.

Even Ciba, it would seem, can't justify its invention sitting in landfill sites forever, leaching triclosan into our waterways. I guess even the chemical industry will acknowledge that bacteria actually belong in some places in this world.

The Germs Bite Back

The over-triclosanitization of the planet wouldn't be such a big deal if it weren't for a few niggling problems:

1. mounting evidence that, in many products, it works no better than competing products that have no triclosan
2. increasing levels in people and the environment that have now been linked to health problems—and the biggie:
3. good reason to think that it's contributing to bacterial resistance, aka the rise of "superbugs"

Let's look at each of these in turn.

First off, are "antibacterial" soaps really no better than "normal" products from competitors? Well, in household settings, this would seem to be the case. Studies published by the American Medical Association, the U.S. Food and Drug Administration, and the Centers for Disease Control and Prevention's journal *Emerging Infectious Diseases* come to similar conclusions: that in household settings, there is no evidence to suggest that the use of antibacterial soap is more beneficial than the use of soap and water in reducing bacteria or the rate of disease.[5] Another study, of two hundred American households, concluded that people who use antibacterial products have no reduced risk for infectious disease symptoms.[6]

Recently, another large study of American households found that soaps containing triclosan at concentrations of less than 0.3% were generally no more effective than plain soap at preventing infectious illness symptoms and reducing bacteria on the hands.[7] Regardless of where the samples were gathered, there was little benefit associated with the use of soap containing triclosan compared to using plain, regular soap.[8]

Stuart Levy, one of the authors of this study, points out that "in household products, triclosan is probably somewhere around one-fifth or one-tenth the concentrations that are used in hospitals." Levy supports the prudent use of triclosan ("it's great in hospitals") but objects to its use in lower concentrations in a more widespread way. Enough chemical is being put in the products to tout them as "antibacterial" but not always enough to actually kill the germs on our hands.

Secondly, evidence of the bioaccumulative and persistent nature of triclosan is mounting. It tends to build up in animal and human fat tissues and has been detected in umbilical cord blood as well. Research in both Europe and the U.S. has found triclosan present in the umbilical cord blood of infants. Swedish studies have documented high levels of triclosan in women's breastmilk.[9] In one paper published in 2002, it was found in three of five breast-milk samples. The Centers for Disease Control and Prevention found triclosan in the urine of 75 per cent of the more than 2,500 Americans tested.[10]

As the use of triclosan becomes more widespread in consumer products, the likelihood of its being emitted into waterways also increases, since approximately 95 per cent of products containing triclosan end up going down the drain. Triclosan was one of the most frequently detected compounds in a U.S. Geological Survey of American streams, likely a result of its presence in discharges of treated wastewater. The survey studied 95 different organic wastewater contaminants.[11] While wastewater treatment can remove much of the triclosan and other compounds, not all compounds will be removed. And in addition to the concerns about triclosan's presence in water, triclosan risks having toxic effects on algae and aquatic ecosystems.

Concerns have also been raised about triclosan's interference with thyroid activity. In a study of mice, it was found that triclosan affected body temperature, lowering it, and caused a depressant effect on the nervous system.[12] Japanese studies of fish have not demonstrated estrogenic activity in fish exposed to triclosan but rather androgenic effects, causing changes in fin length and sex ratios.[13] Studies on frogs have shown that low levels of the chemical can interfere with normal thyroid function, triggering rapid transformation of tadpoles into frogs.[14] In Scandinavia, government officials have discouraged the use of triclosan as a result of possible endocrine disruption as well as potential bacterial resistance.

And finally, the superbugs. The question of whether the prevalence of triclosan is causing bacterial resistance is the hottest debate surrounding this chemical. Some research points to this as a possibility.[15] The American Medical Association went so far as to recommend against the use of antibacterial products in the home, citing evidence of antimicrobial resistance.[16]

Not surprisingly, Brian Sansoni of the Soap and Detergent Association in the U.S. calls the allegations about bacterial resistance a "common suburban myth that floats around mainly due to some active university and academia publicity machines." He points out that the only evidence for bacterial resistance stemming from triclosan exposure comes from the lab and that "there is no real-world evidence linking the use of antibacterial products and their ingredients to antibiotic resistance—none." Sansoni thinks it unfortunate that "continuing to hype this hypothesis" detracts attention from the major contributor to antibiotic resistance, "which is crystal clear: it's the overprescription and the overuse of antibiotic drugs. What we've seen, unfortunately, is both of these scenarios equated in the same breath—you know, antibiotic drugs and antibacterial products. It's like comparing Mount Everest to a molehill."

As one of the primary targets of Brian Sansoni's criticisms, Stuart Levy is careful with his words. "I have said clearly that the use of triclosan is not the primary reason for the bacterial resistance that we face today. It's misuse of antibiotics in humans and animals."

Levy continues, "But that doesn't mean we should complacently say that antibacterials aren't an issue. We should look at it and evaluate it and continue to evaluate, but better than that we should ask the question as to whether there is a benefit to the consumer if in fact there is the threat that it could be harmful." Levy acknowledges that the existing evidence comes from the laboratory and says, "The word I've always used is 'potential.' If we can observe this in the laboratory, it's certainly likely to happen in the outside world. If you use anti-microbials enough, you'll get resistance."

One of Levy's colleagues, Dr. Allison Aiello, a Professor of Epidemiology at the University of Michigan's School of Public Health, agrees that the debate about the possible link between triclosan and antimicrobial resistance is important, but she says, "It's also important to keep the big picture in mind." She is concerned, too, about the rapid spread of triclosan throughout consumer products in the absence of adequate regulation because "if we are going to think about this chemical in general, and discuss its use and its efficacy as well as its risks, we need to put into the equation some of the other research people have done in terms of its fate in the environment and its effect on the body, its toxicity and the like."

Germophobia

West Nile Virus. Bird flu. *Listeria.* SARS. Flesh-eating disease. Methicillin-resistant *Staphylococcus aureus.* So often, it seems, there's a new microbe for people to fear.

And not without reason, notes Dr. Chuck Gerba. Gerba, known as "Dr. Germ" in his popular writings, is a Professor at the University of Arizona and a noted authority on germs and how they spread (one of his more infamous research projects involves measuring how toilets spray a veritable germ geyser into the air when they're flushed without the lid closed). In his book *The Germ Freak's Guide to Outwitting Colds and Flu,* Gerba makes the point that a hundred years ago, infectious disease was the leading cause of death. By 1980 it had fallen to number five, but in the last ten years, it has climbed its way back up to number three.

Gerba feels strongly that we need to take stock of the situation and "reinvent hygiene" for this century. We need to do so for two reasons. First, he says, "the population is a lot more susceptible in terms of serious outcomes. One-third of our population falls into this group and these tend to be the elderly, babies, and compromised people—like cancer chemotherapy patients—and pregnant women. We're an aging population, and common diseases like diarrhea are just a mild inconvenience for most people, but if you're over 65 it can be quite life threatening."

The second reason, Gerba says, is "the lifestyle changes we've undergone" that put us in contact with more germs than ever before: "More of our food supply is being imported from the developing world, which is exposing us to more varieties of pathogens than we've ever seen before. Eighty per cent of us now work indoors in offices, whereas a hundred years ago most of us worked in a farm, on a field and went into town once a week. Today we spend our days in office buildings, in super malls, we have cruise ships that have gone from one hundred people to three thousand people, we have stadiums of an enormous size and basically we're sharing more space with more people than ever before. When you do that, you are sharing more germs, with more people, than ever before."

Gerba uses the example of the pen that arrives in the restaurant on the little credit card tray, noting that it was "touched by one hundred people earlier today, and basically they all left behind a trail of all the infections they have for you to pick up." And sometimes germs are in the unlikeliest of places. "You never want to share a cell phone. Nobody ever cleans them. The germiest spot in the hotel room is the TV remote because nobody ever disinfects it," he says. "All these new pieces of electronic equipment that we share are really just germ transfer points."

Germs are tough and adaptable, and "every time we have a lifestyle change, they take advantage of it," says Gerba. He mentions the SARS virus: "It looks like it came from bats; it got into other animals largely because of the expanding human population and closer contact." Another microbe, the *Legionella* bacterium, likes warm water and in the natural environment would have been a problem only "if you sat in a hot spring." But in a world full of showers, hot tubs and fountains, ready-made artificial habitats, *Legionella* has thrived and presents a real threat to human health.

So what to do? Unlike Stuart Levy, Gerba is not convinced of the evidence linking triclosan to bacterial resistance. He is sure, however, that it's just not necessary in many cases. "I have my concerns about triclosan largely because I don't think it's been proven to be efficacious in everything it's been used in," he says. "I think people shouldn't go overboard with this. . . . You don't have to disinfect everything you come in contact with. I mean, even when you go out in public just washing your hands is a good enough strategy."

Unfortunately for the germ obsessed of today, Gerba's common sense approach is often not taken to heart. And for some unfortunate souls, their germ concern rages out of control on a daily basis. Howard Hughes is the classic example of someone who, later in life, was paralyzed with fears of germs and contamination. More recent icons of the clean obsessed include the TV detective Monk, who in the series named for him is depicted as a rising star in the San Francisco police department who has now been rendered obsessive compulsive by the tragic—and still unsolved—murder of his wife. Monk's psychological disorder causes him to fear virtually everything: germs, heights, crowds and even milk. The host of *Deal or No Deal*, Howie Mandel, is very open about his real-life germophobia and avoids shaking hands with guests on his show. On his website there's a joking photo of his daughter playing soccer while encased in a giant plastic hamster ball with the caption "My daughter, Riley, claims that my germ phobia has gone too far and is infringing on her extra curricular activities."

Jerilyn Ross, head of a prestigious anxiety disorder clinic in Washington, D.C., and President of the Anxiety Disorders Association of America has seen it all.

"I had a patient who was so afraid of germs that she didn't even shower," she says. "She didn't think she could get herself clean enough. She would take six to seven hours to take a shower to wash each strand of hair; but because it was so overwhelming, she could only do it once a month. So needless to say it wasn't about being clean. I had one woman who was so afraid there were germs in her carpet she was vacuuming every day; all of the fibres on her carpet had to be aligned in the same direction. So she could spend an entire day at it. I've had at least two or three patients who will actually clean when they do a laundry. Before they put their laundry in, they will clean the washing machine and the dryer with chemicals to make sure it's clean enough to put the soap and the clothes in. I've had people who will clean their hands so much—50 to 60 times a day—that their hands are constantly red and raw."

Ross goes on and on with examples of patients she's had with severe mysophobia (the clinical name for germ phobia), which Ross tells me is really just a kind of highly specific obsessive-compulsive disorder (OCD).

I asked Ross whether our current societal obsession with getting rid of germs reinforces the behaviour of people who are OCD about being clean. "Well, it's true," she says, "you can't walk out of any place today without having those handy wipes. I've worked out with my trainer and there's the dispenser where you can get the handy wipes right beside us. I think that like anything else, for people who already had either vulnerability or the disease of OCD, then that kind of stuff in some ways makes their fear more acceptable. But it doesn't necessarily make it better or worse."

Ross points out that OCD is a clinical disorder, a disease, so you either have it or you don't. But she uses the example of something she observed after 9/11 to illustrate how social attitudes make it easier for people to talk about their disorders. "After 9/11 people were no longer embarrassed or ashamed to say they were afraid of flying. They would tell their boss, 'I can't go on that trip; I don't want to fly' and the boss would say, 'Yes, yes, I understand; I wouldn't now either.' So it kind of legitimizes something they need help for; same thing with this."

Donald Trump fears germy elevator buttons. Jennifer Lopez is reported to keep her twins' room sterile. Oscar-nominated actor Terrence Howard requires the women he dates to use baby wipes rather than toilet paper in the washroom, because to do otherwise would be unclean.

The germophobes are out of the closet.

The Filthy Truth

We weren't always into being so clean.

In fact, many centuries went by when our ancestors were not only suspicious of bathing but were actually convinced that baths were hazardous to life and limb. They positively revelled in their griminess. According to Katherine Ashenburg, the history of human attitudes to personal hygiene in the West can be organized into a few distinct periods. "The Romans were pretty clean," Ashenburg says. "You would spend a couple hours a day in a Roman bath in the 1st century. Everything fell apart in the Dark Ages. Then the Crusaders went to fight in the Near East and came back with the thing that remained of the Roman bath in the Byzantine Empire—the steam bath. They brought this back in about the 11th century, and our medieval ancestors built great village steam baths and had a great old time until the plague came in the 14th century." As every third person in Europe started dying, one of the culprits fingered by the medical establishment as the carrier of disease was the steam bath and its warm water. In 1348 the medical school at the Sorbonne decided that warm water opens the pores and allows disease to enter. And from then on, for about four to five centuries—depending on where you lived—"people were petrified of warm water and they almost did without it," according to Ashenburg. Baths were looked upon as things to be taken only as desperate remedies: a medical necessity prescribed by doctors. Something conducted under medical supervision only when people were really sick and all else had failed.

Ashenburg's favourite story to illustrate this prevailing attitude occurred in 17th-century France. King Henri IV sent for his Minister of Finance, requesting his presence at the Louvre. When the King's messenger arrived at the Finance Minister's house, he was horrified to find the Minister taking a bath. This was a huge, three-day procedure, often involving enemas and other extreme preparations. The Finance Minister prepared to get out of his bath immediately, but the attendants said, "No, you mustn't!" The messenger then raced back to the Louvre. At this point the King's physician said, "Oh, my god, this is serious. This is a man taking a bath under medical supervision. This is a catastrophe. He mustn't move." The King then sent his Minister a message saying, "You must stay in your bath. Do not abort your bath. I will personally come and visit you tomorrow." The King didn't normally go to his Ministers' residences, but in this case the involvement of a bath was so unusual and so frightening that it warranted this extraordinary measure.

One of the most famous non-bathers in history was King Louis XIV, who was very athletic and lived a long life. He took two baths in his entire life and was reported to have not liked either of them (as his physician wrote in his diary). What the King did, and what everybody down to the lowest social classes did, was to dabble his fingers when he got up in the morning in a little wine and water. He and his subjects occasionally washed their hands and sometimes wiped their faces, but nothing else.

Throughout this long, bath-phobic period of human history, people who were fastidious in their personal hygiene habits were thought to be particularly odd and noteworthy. The well-known 19th-century dandy Beau Brummel scraped himself all over with a brush every day and took baths in milk. "Even today we would find him quite crackers, but not half as much as the early 19th century did," observes Ashenburg. Although he was considered to be exceedingly strange, Brummel was well connected in the upper echelons of society, and his new ideas came to have considerable influence. It was he who started the trend of tailored suits and ties for men that continues to this day, and by the time he died, it was much more fashionable for gentlemen to be clean.

"Napoleon was super clean," says Ashenburg. "He and Josephine both took long, long, long hot baths every morning. For Napoleon they were very much connected with stress. So that when the peace of Amiens was breaking down around 1805, he would stay in his bath for a couple of hours and have his aides read him newspapers and telegrams and letters. At one point he was there up to five or six hours a day. He was unusually fastidious for the day."

The target of our modern-day cleaning obsession—the micro-organism—was not even widely recognized as the cause of disease until the turn of the 20th century. For most of human history, people thought that disease was spontaneously generated or was caused by a noxious form of bad air called "miasma." The miasmatic theory of disease became popular in the Middle Ages and was still being defended in the late 19th century by people as prominent as the Crimean War nurse Florence Nightingale. People didn't worry about germs. They worried about nasty smells from things like rotting meat, garbage and putrefaction. In order to combat these, they were obsessed with ensuring good ventilation. It wasn't until scientists like Joseph Lister (after whom Listerine and *Listeria* are named) conducted experiments demonstrating the dramatic benefits of antiseptic measures—such as the reduction of hospital infections by hand washing—that people actually believed germs were real.

In the late 1800s soap became cheap enough that the middle class could afford it. Until that point most people had soap for washing clothes and floors but not bodies. And with this new refinement, which included better body soap, broad-scale advertising was born. As Ashenburg explains in a chapter of her book called "Soap Opera,"[17] soap and advertising grew up together. "Advertising by fear"—which targeted the insecurities of the average person—quickly became a staple of the industry. Listerine advertisements in the early 20th century claimed that halitosis was a nationwide epidemic and suggested that bad breath would inevitably upset the natural bond between a mother and child: "Are you unpopular with your own children?" And unmarried women were targeted with lines like "Skin that says, 'I do!'" or "Till BREATH do us part."

Ads were aimed not only at women, but also at men. The Cleanliness Institute, an organization supported by the majority of soap manufacturers in North America, ran an ad in the late 1920s with the line "There's self-respect in SOAP and WATER" and a graphic of a well-dressed man with a briefcase looking down at an unshaven, dishevelled man. The recent marketing of "antibacterial" products is simply the continuation of a long and beautiful relationship.

Rather than buying into the common notion that our ancestors didn't care about cleanliness because they simply didn't have access to the right technology or plumbing or water delivery systems, Ashenburg flips this argument on its head. She believes that technology follows desires. "The Romans had technology for water delivery and plumbing and heating in their imperial baths. That knowledge was not really lost, but until the 19th century, people weren't interested in it. The English, who were more interested in being clean in the 19th century than the French, were able to have plumbing in the majority of London houses by the 1830s. The French knew about this but declined to follow suit. All of this was really about mentality. The French didn't care to be clean in the 1830s and for some complicated historical and sociological reasons, the English cared more." Ashenburg's conclusion is that people could have been clean in lots of countries, centuries earlier, but it was not a matter of interest to them. "I think our relationship to how clean we want to be has all kinds of things to do with our religion, our relationship to our own body, our feelings about privacy, individuality, sexuality, all kinds of things. Cultures that believed more in a communal sense were much less bothered by the fact that they smelled or their neighbours smelled."

The current unparalleled Western obsession with hygiene—"pretending that we're not of this Earth" as Ashenburg says—reflects some very recent changes in societal desires. She concludes that our "germ craziness and all these antibacterial things" are connected to fears like "terrorism and 9/11. Germs, like terrorists, are unseen enemies, and you never know when they're going to strike. I think a lot of the current hygiene thinking is about the American wish to control things."

The irony being, of course, that the current rate of antibacterial use has unleashed a wave of triclosan on the population in an entirely uncontrolled and largely unmonitored manner.

Pesticide Toothpaste

Looking at the array of highly scented, luxuriously packaged, triclosan-infused toiletry items I assembled for our experiment—surely the pinnacle of soap evolution—humanity's former long-standing disregard for bathing and tolerance of stinkiness seemed very remote indeed.

As opposed to the other chemicals dealt with in this book, our triclosan experiment was comparatively easy to organize. Triclosan is well labelled on products, and I had been actively avoiding it for years. If "antibacterial" appeared anywhere on a container, the product didn't make it into our house. But for purposes of our experiment, I purchased a variety of off-the-shelf products containing triclosan at local grocery stores and used them in a normal way over a two-day period. I felt a little strange deliberately exposing myself to triclosan, because unlike phthalates and bisphenol A—which stay in the body for only a few hours—triclosan sticks around for several days. It would take over a week for my body to rid itself of the chemical.

In preparing our experiment we had looked at a few studies of triclosan to try to gauge what to expect. Researchers using laboratory preparations of the chemical in skin cream and mouth rinses had demonstrated how easy it was to increase levels in the urine through these single sources.[18] But there are now so many triclosan products on the market that people can easily be exposed to multiple triclosan sources simultaneously. A survey of consumer products in 2000 found that over 75 per cent of liquid soaps and nearly 30 per cent of bar soaps (45 per cent of all the soaps on the market) contained some type of antibacterial agent. Triclosan was the most common agent found. In Sweden, for example, in 1998 alone, 25 per cent of the total amount of toothpaste sold contained triclosan.[19]

Table 5. Rick's triclosan shopping list

Bathroom:
Colgate Total toothpaste
Clean & Clear foaming facial cleanser
Dial Complete triclosan handsoap
Gillette shave gel
Right Guard deodorant
Dettol pine fragrance shower soap

Kitchen:
Dawn Ultra Concentrated dish liquid/antibacterial
  handsoap
J Cloth (apple blossom scent with Microban)

After using these products for a mere two days, the increase in triclosan in my body was simply stunning. My urine levels of triclosan went from 2.47 nanograms per millilitre (ng/mL) before exposure to 7,180 ng/mL after exposure. The difference between these numbers is hard to depict visually: Because of the huge increase, you can barely see my starting value on the graph. And why, given the fact that we've banished triclosan from our home for years, wasn't my starting value zero? Likely because of the levels of triclosan that are now found in water and food that we all absorb day in and day out.[20]

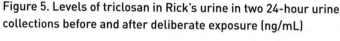

Figure 5. Levels of triclosan in Rick's urine in two 24-hour urine collections before and after deliberate exposure (ng/mL)

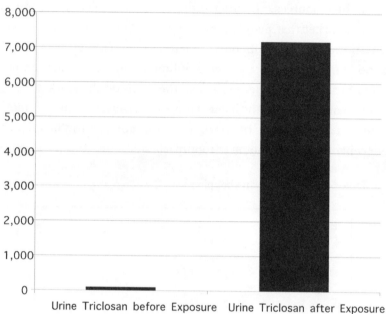

Urine Triclosan before Exposure   Urine Triclosan after Exposure

My results were very interesting when compared with recent testing by the U.S. Centers for Disease Control and Prevention (CDC). In 2003 and 2004 an analysis of the triclosan levels in the urine of over two thousand Americans revealed that the geometric mean was 13 ng/mL, with a range in values from 2.3 to 3,790 ng/mL.[21]

So after two days, my self-experimentation took me from the very bottom of the heap to far above the highest value recorded to date in the U.S. population. To accomplish this, I used eight different products containing triclosan all at once, a number not likely used by many people. But judging from the hundreds of nanograms of triclosan in the urine of some of the CDC test subjects, the simultaneous use of a few triclosan products is not uncommon.

When I told Klaus Nussbaum from Ciba about the triclosan levels in my urine, he said, "It shows that your body is working very properly in removing triclosan." When I asked if 7,180 ng/mL was at all cause for concern, he answered, "It's high for the moment, your body is working properly, but you should know that the human body usually adapts in the metabolism of triclosan." I was struck by Nussbaum's use of the word "adapt." *The Oxford Dictionary of Science* defines adaptation as "any change in the structure or functioning of an organism that makes it better suited to its environment." The concept that our bodies need to adapt to synthetic chemicals is an interesting one, and biologically true. Given that triclosan is a human creation, the metabolic pathways necessary to break it down and excrete it are indeed things that our bodies need to learn how to do.

Early on in the writing of this book, I was interviewed by a film-maker who was making a documentary about the impact of toxic chemicals on human health. At one point in our interview, I was asking him more questions than he was asking me. The huge number of synthetic chemicals that surround us every day of our lives is "creating a new kind of evolutionary pressure," he believed—a new kind of natural selection every bit as powerful as the process that resulted in human populations developing lighter or darker skin pigments in response to prevailing climactic condi-tions. Because of the increasing evidence that many human ill-nesses—including fatal ones like cancer—are linked to exposure to chemical pollution and because some people's bodies are better able to adapt to, and cope with, these new environmental stressors than others, this filmmaker wondered aloud whether the human population is "being culled" by toxic chemicals. Because of the luck of the biological draw, some of us are genetically predisposed to deal with the daily assault of toxic chemicals and some are not.

I don't know whether the future form of humans is being deter-mined by the chemical soup we're living in—or whether the defi-nition of our evolution as a species has now changed from "Survival of the Fittest" to "Survival of the Chemically Immune." What I do know is that the ubiquity of synthetic chemicals like triclosan in our daily lives and their accumulation in our bodies is an unjustifiable imposition by the industries who manufacture the chemicals and by the government regulators who are supposed to keep such things from causing harm.

And I certainly don't want any new metabolic pathways for triclosan being activated in my body without my consent.

Nano and the Toxic Treadmill

As if triclosan wasn't a big enough problem, the latest incarnation of the runaway antibacterial train presents an even more complicated challenge: nanotechnology, the creation of super-small particles that are only a few atoms or molecules in size.

In recent years—with no fanfare and largely unknown to consumers—brand-new kinds of nanotechnology have become commonplace. According to the Project on Emerging Nanotechnologies, an inventory of nanotechnology-based consumer products, the number of products has increased threefold between March 2006, when the inventory was released with 212 products on the market, and August 2008, when there were 609 products. Of the emerging nanotechnologies, nanosilver is most commonly used in consumer products,[22] with over 200 manufacturer-identified nanosilver products on the market.

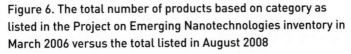

Figure 6. The total number of products based on category as listed in the Project on Emerging Nanotechnologies inventory in March 2006 versus the total listed in August 2008

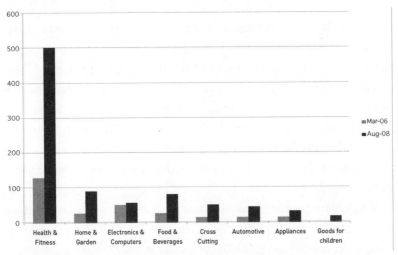

According to Dr. Andrew Maynard, Chief Science Advisor to the Project on Emerging Nanotechnologies in Washington, D.C., average consumers will regularly encounter these nanomaterials and not even know it: "There's quite a range of nanomaterials at present," says Maynard. "People will come directly in contact with things like nano-sized silver particles that are used as antibacterial agents. They'll come in contact with nano-sized titanium dioxide and zinc oxide in cosmetics and sunscreens. They will come across some nano-encapsulates that are used in some cosmetics. They will also, indirectly, come in contact with things like cerium oxides being used as a fuel additive and carbon nanotubes that are usually embedded in a product; so people aren't going to be directly exposed to them, but they'll be using products that contain them."

And what's the point of nanosilver? As Maynard explains, silver has been known as an effective antimicrobial agent for centuries, but it's not that easy to use. It's very difficult to put a large lump of silver into a product. And the only other available option is to use it in chemical form, such as silver ions. Over the last few years, manufacturers have discovered that if you melt silver into very small particles—about 20 or so nanometres in diameter—you can effectively incorporate those particles into a wide range of products, thereby giving them some degree of antimicrobial capability. "So now you're seeing nano-sized silver particles appearing in things like surface coatings, clothing like socks, surfaces of food containers and refrigerators. Almost any product where you can see this foreseeable market for antimicrobials," says Maynard. "The market for antibacterial silver products is projected by some market analysts to grow to 110–230 tonnes of silver per year in the 25-member European Union by 2010; a similar-sized market might be expected in the USA."[23]

While nanoparticles may be new, our scientific understanding of silver is not. The EPA classifies silver as an environmental hazard because of its toxic effects on aquatic plants and animals. A study published in 2005 found that nanosilver is 45 times more toxic than regular, standard silver.[24] Another study found that nanosilver has the potential to destroy beneficial bacteria used in wastewater treatment.[25] According to a paper published in September 2008, nearly one-third of nanosilver products on the market in September 2007 have the potential to disperse nanosilver into the environment.

In 2006 Samsung introduced a SilverCare washer that released silver ions into the wash. These ions were then released into the waste stream with each load of laundry. Even when socks containing nanosilver are washed, nanosilver is released into the water discharged from the laundry and eventually makes its way into watercourses. The Stockholm Water Authority claimed that households in Sweden using the nanosilver washing machine would emit two to three times more silver than would be emitted without the use of the washing machine.[26]

In another recent study, the first of its kind, researchers experimented with six different pairs of socks, all of which had been marketed as anti-odour and impregnated with nanosilver. They found that the socks released varying amounts of silver. Some released silver after the first wash, other socks gradually released the silver after multiple washes and others released no silver. More research is needed to find out just how much of the silver particles make their way from the sock into the washing water and ultimately into wastewater and the broader environment, including aquatic life and humans. But meanwhile, for the average consumer standing in front of a display of nanosilver socks in a store, there is virtually no way of knowing which socks will release silver and which will not.

Given the lack of monitoring of nanosilver and what Andrew Maynard calls the "very complex" behaviour of nanosilver particles in the environment, more research in particular is needed concerning the impact of nanosilver on soils. Bacteria, after all, are what make soils work. So having soils peppered with a potent, "space-age," antibacterial agent is a bit of a problem. What few studies exist suggest that nanosilver is toxic to bacteria that consume inorganic material and thus release crucial nutrients that are essential to the formation of soil.[27] Toxicity can also affect a bacteria-driven process known as "denitrification," in which nitrates are converted to nitrogen gas in some soils, wetland and other wet environments. This process is critical because excess nitrates reduce plant productivity. They can also result in "eutrophication in rivers, lakes and marine ecosystems, and are a drinking water pollutant."[28] Nanosilver's toxicity has also been demonstrated in studies that show its effects on mammalian liver cells, stem cells and even brain cells.[29]

I went to an evening panel discussion on nanotechnology recently in Toronto. It featured some experts in the field and focused on the prospects for properly regulating this new, and exploding, class of products. I must admit that in my darker moments that evening, listening to the extent to which nanomaterials are entirely unregulated and learning that even many manufacturers admit they do not fully understand what they're dealing with, I was struck with a feeling that we're on a sort of Toxic Treadmill. No sooner do we deal with one chemical that's harming our health than we see another one coming along. We can't get off the treadmill and never seem to learn from our mistakes.

I put it to Andrew Maynard that nanosilver is a classic example of this phenomenon, but he was actually more positive than I expected. "It's good that we're still at the starting stages of different types of nanotechnologies being developed and already we're having a fairly broad debate about how you would bring these technologies forward responsibly," he says. "Even though things seem to be a little bit dicey, we're doing a lot better with this technology than we have with previous ones."

The Merchants of Fear

The bottom line is that there's a little Howard Hughes in all of us, and the chemical industry preys on this big time.

The avalanche of advertising extolling the (in many cases) nonexistent virtues of triclosan and fomenting a society-wide germ panic is the prime example of this, but the use of fear to peddle chemicals is a theme you'll find in other chapters of this book as well. Fear of fire sells more flame retardants. Fear of insects sells more pesticides. Fear of the odours of daily life sells phthalates. Recent industry campaigns to defend the unrestricted use of phthalates and bisphenol A have variously suggested that children will be deprived of their toys and that huge numbers of crucial products will be yanked overnight from grocery store shelves.

One of the most egregious examples of this approach that I have seen are the tactics used by Procter & Gamble to sell more bottles of its triclosan-containing Vicks Early Defense Foaming Hand Sanitizer. The sophisticated Vicks website is aimed squarely at mothers. "No one knows the power of Mom like Vicks does. We know you have the power to help create Germ-Defense Zones wherever you go—at home, in your car, at work, or at your child's school. So go ahead, take the pledge and then enlist friends, family, and neighbors to help create their own Germ-Defense Zones." The website has a function that allows moms to take "Vicks Mom Challenge: The Pledge," which goes like this: "I pledge to help create a Germ-Defense Zone wherever possible. I've read the tips provided by Vicks, and I will use this knowledge to try to stop the spread of germs. In addition, I plan to use Vicks Early Defense Foaming Hand Sanitizer to further help me fight the good fight." Those taking the pledge enter their zip codes and can see their registration pop up on a map of the United States: a snazzy visual representation of the march of "Germ-Defense Zones" across the land.

If this isn't marketing through fear, I don't know what is. In September 2007 the U.S. Food and Drug Administration ordered Vicks to stop advertising its Early Defense product because of its erroneous claims that it can help prevent the spread of viruses that cause colds. A Vicks spokesman was quoted at the time as saying that the company believed it was operating within FDA guidelines and that it was going to work with the FDA to clear up any misunderstandings. As of this writing the product was still being marketed throughout the United States, though no longer with the same claim.

The irony in all this is that environmental advocates, including yours truly, are frequently accused of "fearmongering" by the chemical industry. "Our products are safe; don't listen to the scare tactics suggesting otherwise" goes their refrain. In reference to our Toxic Nation campaign, I was once called the "master of linking discombobulated facts with alarming suggestion and exaggeration." In a recent meeting an industry representative actually called me a "chemophobe," which I thought was a cool word, even though I disagreed with the guy's premise.

So let me put it on the table: I am not a chemophobe. I love chemicals. Most of the chemicals in my daily life, including caffeine and alcohol and even the low-VOC paint on the walls of my home, are just dandy.

In my third year of university, I tried my hand at a summer of tree planting in the wilds of British Columbia. I wasn't terribly good at it and didn't enlist a second year. It was tough work. Piece work. You were paid based on your speed and agility and the number of trees you rammed into the ground. I remember one day the bugs were so bad I thought I was going to lose my mind or scratch my own eyes out or both. I pulled my long-sleeved shirt over my head so my face was framed by the neck hole and I doused myself—including my head—with a full bottle of the strongest, DEET-ridden, insect repellent imaginable. DEET is a chemical so strong it actually melts plastic, and you can't even buy bug dope now in the concentrations I was using 20 years ago. But there was no other option on the cut block that day.

So here's a message to the chemical industry: I'm no chemo-phobe. I'm downright chemophilic in some circumstances. What I object to are the chemicals like triclosan that aren't necessary, are possibly dangerous and are foisted on us every day without our knowledge or consent.

SEVEN: RISKY BUSINESS: 2,4-D AND THE SOUND OF SCIENCE

[IN WHICH BRUCE PONDERS PERVASIVE PESTICIDES]

A lawn is nature under totalitarian rule.

—MICHAEL POLLAN

HEALTHY GREEN LAWNS are lovely to look at and great to lie on. It's no wonder they're the object of envy in the famous saying "The grass is always greener on the other side." I have vivid memories of mowing, watering, fertilizing and chemically treating my family's lawn as a teenager. There was great satisfaction in seeing a freshly cut, weedless, green lawn with clean diagonal mower tracks. And maybe even some pride in having people think that the grass truly was greener on our side of the fence.

Green lawns require work. They are a sign of care and dedication and, in some contexts, they're even evidence of being a good neighbour. But we have to ask ourselves, "What are our priorities?" Are weed-free lawns worth the risk of children suffering from respiratory disease? Or lawn-care workers getting non-Hodgkin's lymphoma? Or future generations of kids suffering from learning disabilities? I don't think anybody using lawn-care products wants any of that. In fact, most of us are unaware of the potential dangers of pesticides.

The truth of the matter is that lawns are giant pesticide guzzlers. They consume 90 million pounds of pesticides and herbicides each year in the United States.[1] In a strange coincidence this is also the amount of chicken wings consumed by Americans on Super Bowl weekend, and though the toxicity of chicken wings may be open to debate, pesticides are definitely poison.[2] That's the whole point. They kill pests. Insecticides kill bugs and herbicides kill weeds. Lawn and garden fertilizers that control weeds, known as "weed and feed" products, contain the chemical 2,4-D, the most widely used herbicide in the world.[3]

"DDT Is Good for Me-e-e!"

Most sayings have ancient roots, but "The grass is always greener on the other side of the fence" is strangely contemporary. Its first recorded use was in 1957.[4] This also happens to be the year that DDT spraying near waterways was banned by the U.S. Forest Service.[5] And it's the same year that the *New York Times* reported on the failed attempt of Nassau County, New York, to ban DDT. This story, in turn, led to the editor of the *New Yorker* convincing Rachel Carson to begin writing *Silent Spring*, her pioneering work on the damaging effects of pesticide spraying.[6] The late 1950s were the heyday of DDT in North America, with total use peaking in 1959. These may all be strange coincidences. But is it possible that chemical companies, seeing the demise of DDT and the future of 2,4-D, invented the expression "The grass is greener . . ."? Or does this sound like pesticide paranoia?

Extensive pesticide use had begun not that many years before—after the end of World War II, which had brought large-scale chemical pesticide manufacturing to the United States. Mosquito-borne diseases, mainly malaria and typhus, were wreaking havoc on the troops in southern Europe, northern Africa and Asia. The military was eager to find a solution. Synthetic pesticides were not in wide use, or even well understood, for that matter, and DDT did not even exist prior to World War II. But between 1943 and 1944, military demand for DDT shot up from 10,000 pounds a month to 1.7 million pounds a month.[7] Desperate for DDT the

U.S. government provided 100 per cent tax write-offs on the construction of DDT-manufacturing plants and forced Geigy, the DDT patent owner, to give DuPont a licence to produce DDT, even permitting sales after the conclusion of the war.[8]

After the war American companies were left with huge DDT production capabilities but no market. Manufacturers were well aware of the economic potential for DDT, and in some ways they viewed the demand for the chemical during the war as simply a means for them to develop government-subsidized production capacity. Despite concerns raised by scientists, DDT became an overnight sensation, and American farming rapidly shifted to the chemical-input model of today. DDT was also used to eradicate garden pests and houseflies, and its huge success led to the invention of all kinds of synthetic chemicals for killing bugs and weeds. Chlordane, dieldrin and aldrin are three chemical relatives of DDT created in the 1940s to target various insects, such as termites, moths and grasshoppers. The patent for 2,4-D was issued in 1945.[9] As one of the world's first "hormone herbicides," 2,4-D "laid the corner stone of present-day weed science."[10]

Though synthetic pesticides were the talk of the town and part of the postwar utopian view of a world with plentiful food and no disease, the potential perils of DDT use were recognized early on. Scientists began to express concern regarding the human health and biological hazards of the chemical in the mid-1940s, calling DDT "the atomic bomb of the insect world" with "possibilities for evil as well as what seems to the human race good."[11] As far back as 1949, the U.S. Food and Drug Administration Commissioner, Paul B. Dunbar, was worried that people exposed to small amounts of DDT and other chemicals over long periods of time might have been in greater jeopardy than soldiers who were briefly exposed—an early recognition of risks to the general population of ongoing exposure.[12]

Military applications occurred in a risk context that was very different from that of the new users of DDT, including American farmers and housewives. Compared with malaria, mustard gas or a bomb, the long-term health hazards of DDT were hardly a consideration. But when the threats were houseflies, gypsy moths or corn weevils, widespread DDT use in homes and on the nation's food supply was open to question. The manufacturers of DDT went to great lengths to sell the benefits of DDT to the American public, as witnessed by the 1947 *Time* advertisement "DDT is Good for Me-e-e!" Creating "meatier" beef, "healthier" homes and apples with no "unsightly worms," DDT was proffered as the solution to many problems. But by 1949 the bloom was coming off the rose. All of a sudden DDT began to lose some of its insecticidal effect. Mosquitoes became resistant and required ten times the dose before they would die. And as its intended victims became more immune to its killing power, the impact on DDT's unintended victims became impossible to deny.

For a "miracle" product, DDT's days were short lived. In 1972, less than 30 years after its first commercial application, all uses of DDT were banned in the United States and in many other countries. Two years later the U.S. banned the use of DDT's toxic cousins aldrin and dieldrin. Unused, yes. Gone, no. Decades later DDT still exists at measurable levels in the environment, and its persistence ensures that its toxic legacy will continue for the foreseeable future. (DDT use continues in countries where malaria is prevalent.)

According to a 2008 study, men with DDE, a byproduct of DDT, in their bodies are 1.7 times more likely than those without DDE to develop testicular cancer.[13] New studies show that DDT compounds contribute to breast cancer development by blocking the actions of natural hormones that slow down the growth of cancerous tumours.[14] It seems incredible that DDT was used widely in North America for only three decades but can still be causing cancer nearly 40 years after it was banned. This is a powerful lesson in the dangers of highly persistent toxic substances.

Ad for DDT in *Time* magazine, June 30, 1947

2, 4-D, the Hormone Herbicide—and Me

As DDT's fortunes waned, those of 2,4-D waxed.

2,4-D, short for 2,4 dichlorophenoxyacetic acid, is a synthetic chemical herbicide. More importantly, it is one of the earliest "hormone herbicides." Working its magic by disrupting a number of hormone processes in plants, 2,4-D causes them to grow uncontrollably and keel over dead. It was designed primarily to kill broadleaf weeds (think dandelions), weedy trees and aquatic weeds (seaweed that gets in the way of oyster farming, for example). It is especially valued because it kills selectively, targeting flowering plants and trees but sparing grasses and their relatives. That is why we can spread 2,4-D all over our lawn and kill the weeds but not the grass. The popularity of this chemical among farmers stemmed from the fact that corn, grains and rice are in the grass family, making 2,4-D the perfect chemical to kill weeds and plants that grow between rows of these crops.

Like many pesticides 2,4-D is associated with a number of potentially serious health hazards for humans. In fact, the list of known or suspected health effects reads like an inventory of the worst possible things that could happen to a human. And I'm not even referring to things like the nausea, headaches, vomiting, eye irritation, difficulty breathing and lack of coordination that can occur from accidentally spilling 2,4-D on your skin. I'm referring to the long-term effects of exposure to 2,4-D: non-Hodgkin's lymphoma (a form of blood cancer), neurological impairment, asthma, immune system suppression, reproductive problems and birth defects.[15]

This pesticide also has special notoriety because it was one of the active ingredients in Agent Orange, the chemical spray used in the Vietnam War to clear jungle foliage. Agent Orange, which also contains a number of even deadlier ingredients, is at the centre of ongoing medical and legal battles initiated by soldiers seeking compensation for the cancers and other ailments they attribute to their wartime exposure.

Given all these deleterious side effects, Rick and I weren't enthused at the prospect of experimenting with 2,4-D, but low-level, short-term exposure is at least less damaging than long-term exposure. So we thought long and hard about the best experiment to increase, and then measure, 2,4-D in my blood. An obvious test was for me to spray some 2,4-D–laced herbicide on somebody's lawn and measure the 2,4-D in my blood before, during and after the spraying. Remarkably, had we been writing this book a year earlier, this would have been a fairly simple activity to organize. When we first started to plan our chemical exposure experiments, we knew we could not do the 2,4-D test in Toronto, where we live, because the city had banned the cosmetic use of pesticides starting in April 2004. Even so, and even though there are nearly two hundred municipal bylaws in Canada banning pesticide use, we figured we could still find a few pesticide-friendly municipalities nearby. This, of course, raised ethical issues about deliberately poisoning someone else's yard so we could contaminate ourselves.

Just as we started laying our plans, amazingly, the Government of Ontario decided to ban the cosmetic use of lawn herbicides and pesticides throughout the province. So our test was about to become illegal in Ontario! We figured we could still have squeaked in under the wire and found a location near Toronto, but something didn't feel quite right about going to the suburbs to spray a toxic chemical on someone's lawn when we knew the chemical was about to be banned there.

In the end we decided against trying to measure an increase in my 2,4-D levels. Instead, we did a one-time test for a variety of pesticides in my blood—the same testing that Environmental Defence conducted on Canadian families in 2007 as part of the Toxic Nation study, which helped bring the issue of toxic chemicals to the forefront of public consciousness in Canada.[16] Because about 50 per cent of my diet is organic, and it's been shown that people with organic diets, especially children, have lower pesticide levels in their bodies, I assumed that my results would be pretty clean.[17]

There was good news and bad news as it turned out. The bad news was that—like every other person tested by Environmental Defence—I had pesticides coursing through my veins that had been banned for years. Hexachlorobenzene, a fungicide used mostly on grain, was phased out in Canada and the United States by the early 1970s, but in 2008 it was still present in me at a level of 1.2 ng/mL (nanograms per millilitre). This was somewhat higher than the U.S.

average.[18] At 2.9 ng/mL my DDE levels were very similar to the U.S. mean of 3.5 ng/mL.[19] Though banned in the United States and Canada since I was a young child, this telltale sign of DDT pollution still lingered in my bloodstream. Chlordane, a pesticide commonly used on crops like corn and citrus fruits and on home lawns and gardens right up to the late 1980s, was detectable in me in the form of two breakdown products: oxychlordane and trans-nonachlor. Finally, traces of the agricultural pesticide and louse treatment Lindane, still in use in North America until recently, were also found in my body. In addition to being in me, all of these pesticides are found in a large percentage of the U.S. population.[20] [21]

Table 6. Bruce's pesticide levels

Pesticide	Bruce's Pesticide Levels (ng/mL)*	Health Effects
Hexachlorobenzene (HCB)	1.2	Recognized: carcinogen, reproductive/developmental toxin Suspected: hormone disruptor
Beta-BHC (Lindane)	0.5	Recognized: carcinogen, neurotoxin Suspected: hormone disruptor, reproductive toxin
Oxychlordane (Chlordane)	0.4	Suspected: hormone disruptor
Trans-nonachlor (Chlordane)	1	Suspected: hormone disruptor
DDE	2.9	Recognized: carcinogen, reproductive/developmental toxin Suspected: hormone disruptor, respiratory toxin

* Measured in serum

The good news was that interestingly, 2,4-D was not found in me at detectable levels. There are a couple of possible reasons for this. 2,4-D is water soluble, not fat soluble like many other pesticides, so it does not accumulate in the body's fatty tissue. In addition, your average 2,4-D molecule has a half-life in the environment of a few months at most, and Toronto's ban on the stuff had been in place for over four years at the time my blood was tested. Many of my fellow North Americans have not been so fortunate. In a 2005 report, for instance, the U.S. Centers for Disease Control measured 2,4-D in a subsample of the U.S. population between the ages of 6 and 59 and found that one-quarter of Americans who had their blood tested in 2001 or 2002 had detectable levels of 2,4-D in their bodies.[22] A recent study looked at 2,4-D exposure among 135 preschool-aged children and their adult caregivers in 135 homes in North Carolina and Ohio.[23] (Neither state had banned 2,4-D.) Samples were collected over a 48-hour period from solid food, liquid food, hand wipe and spot urine samples as well as environmental samples. The study found that median urinary 2,4-D concentration was more than twice as high in children from Ohio as in children from North Carolina (1.2 ng/mL in Ohio versus 0.5 ng/mL in North Carolina). However, the median concentration was identical, at 0.7 ng/mL, for both North Carolina and Ohio adults.[24] In addition, the pesticide was apparently pervasive in the areas studied in these two states, as 2,4-D was found in more than 80 per cent of the dust samples in all the homes of study participants.

The study authors cannot explain the discrepancy between the 2,4-D levels of children in the two states other than to suggest that children in Ohio may be exposed to higher levels of 2,4-D in things other than their food. Homes in Ohio, for example, had three times more 2,4-D in carpet dust than homes in North Carolina, and kids' hands in Ohio had five times as much 2,4-D residue as kids' hands in North Carolina. However, the real issue is not so much the relative amounts of 2,4-D found in the children but the fact that the researchers found any of the possible cancer causer at all.

Spray, Baby, Spray

Despite the massive growth in organic food over the past decade, pesticides continue to be used in extraordinary quantities.

According to the U.S. Environmental Protection Agency, world pesticide expenditures totalled more than $32 billion in 2000. Expenditures on herbicides accounted for the largest portion of total expenditures, at more than $12 billion.[25] And Americans spread over 9 million pounds of 2,4-D on their lawns every year.[26]

One of the most powerful challenges to environmental progress is inertia. The chemical industry works hard to maintain the inertia of continued pesticide use and thus protect the status quo. The entire procedure works like a game, but the rules are written by the companies that manufacture pesticides. The winners sell more chemicals, and the losers may have much more at stake: cancer and neurodevelopmental disease, for example. The rules of the game are simple. First is the notion that chemicals currently in use are safe, *de facto*, so companies fight hard to keep existing products on the shelf and product bans are almost always followed by lawsuits initiated by the manufacturer. As a result governments are much less likely to restrict or ban an existing product than a new product. In Canada the onus is on public health advocates to prove that a product is harmful; therefore, the historical preference has been to restrict products only after they have been banned in other countries. Canadians live with some of the most lax pesticide standards of any Western country. So we can't assume governments are looking after our health. Thankfully, there are signs that the Canadian government may be starting to set aside its traditional timidity as it stakes out a global leadership position on substances such as bisphenol A (see Chapter 8).

The second rule of the game is never to question the fundamental purpose of a product and certainly never to introduce alternative means of reaching the same end. Take lawn care and 2,4-D for example. The perceived need for smooth, green, weed-free lawns is essential to pushing 2,4-D in the market. Many scientists who do studies financed by the pesticide industry intentionally or unintentionally reinforce the status quo. For example, in a study on environmental persistence and human exposure to 2,4-D conducted by the University of Guelph's Centre for Toxicology, the study begins with the line "Pesticide use can be an important component in well designed programs to maintain turf grass. . . ."[27] This study could also have started with the line "Turf grass is a largely nonessential ground cover requiring the application of toxic pesticides."

Lawns created with native grasses and shrubs do not require pesticides and can be much easier to maintain than a conventional grass lawn. However, not too long ago some municipalities tried to ban alternative forms of lawns. The famous Doug Counter case in Toronto, in which Environmental Defence was heavily involved, is one example. In a regressive move, predating its pesticide bylaw banning the cosmetic use of pesticides on lawns, Toronto City Council was intent on mowing down an avid gardener's native plantings, but this case was successfully defended. In another case a woman in Manitoba was taken to court by her municipality over her native-grass lawn, which was considered to be unkempt. Environmental lawyers took on her case and her interest was also successfully defended.

Happily, better lawn management has now progressed at a rapid pace, and municipalities in Canada now lead the world in their efforts to restrict pesticide use on lawns. (In Manitoba, for instance, the two largest cities have adopted cosmetic pesticide bylaws.) What has now become a substantial grassroots (no pun intended) citizens' movement to ban lawn pesticides in hundreds of municipalities across the country first started in the quaint town of Hudson, Quebec, on the outskirts of Montreal. In 1991 Hudson passed a bylaw restricting the cosmetic use of lawn pesticides. This was an innovative move, the first of its kind in Canada. Predictably, a number of chemical companies took the little town to court, claiming that it did not have the legal jurisdiction to ban pesticides. They argued that pesticide controls required provincial or federal regulation. (It is noteworthy that the safety of pesticides was not their primary point.) Municipalities, they thought, were not allowed to regulate the use of pesticides within their boundaries. In other words somebody was trying to change the rules of their game, and this did not sit well with the companies involved.

A decade passed before the Hudson lawsuit was settled after going all the way to Canada's Supreme Court. But in 2001 the town won the legal right to maintain its bylaw. This paved the way for the dramatic, and rapid, passage of municipal pesticide bans across Canada. What's more, the Supreme Court case was won in part on the precautionary principle. That is, although it was not definitively proven that pesticides on lawns cause specific cancers in people, it was known that many pesticides are linked to health problems, and as the Court ruled,[28] the "weight of evidence" leaned in favour of taking preventive or precautionary action to protect human health and therefore to support the pesticide ban.

The Hudson decision was a critical precedent behind the passage of the City of Toronto's 2004 pesticide bylaw. After Hudson, recounts Gord Perks, a long-time environmental activist formerly with the Toronto Environmental Alliance (TEA), "we went in and fought and fought and fought." Gord is well known in the environmental community in Toronto. He is as articulate as he is radical, and this disarming combination has made him a force to be reckoned with. TEA was the main organization promoting a pesticide ban in Toronto. "For about ten years we slowly built the case up and found allies in nurses, pediatricians and others," Gord recalls—although TEA was the main organization promoting the pesticide ban.

"How did you keep the momentum going?" I asked.

"We used interim battles to establish the arguments and to keep the issue moving. Frankly, the public health department got nervous at times but . . . at a certain point Sheila [Basrur][29] just said, 'No, we're going to do this.' And when it became clear that it was going to come to a showdown at City Hall, a couple of things happened. First of all the industry switched tactics and decided that they would create an appealing—but false—front." Gord was referring to the Toronto Environmental Coalition (not Alliance). It seemed that the pesticide industry set up the coalition as a not-so-subtle way of making their arguments more palatable. They chose a name that sounded grassrootsy and akin to a community organization as opposed to being an industry lobby. The name was also quite similar to that of TEA, which really was a community group. "They were trying to undermine us," Gord said emphatically. "But we were clearly winning the hearts-and-minds battle."

"The Mayor [David Miller] set up a series of meetings between us and the pesticide industry. We thought we had a deal, but a woman from a particular company broke the deal the day before the [City Council] vote." According to Gord all Hell broke loose that weekend, and the industry launched "a massive ad campaign. . . . I tallied it up and I guessed they spent about a million dollars. It was the Dark Ages and they were generating faxes, not emails. And councillors would get a stack several inches high of these faxes from clients of lawn-care companies. And we were, in our little environmental-group way, trying to generate countercalls.

"I remember talking to one city councillor and she said she wasn't switching her vote because she had all these constituents telling her, 'Don't ban pesticides.'" Gord got her to agree that if he could drum up more than 75 voters who supported the pesticide ban, she'd vote for it. "We went to soccer games and arenas and door to door . . . and we came back with 150 people. And she voted our way." That just left the Mayor. He backed the deal he helped broker and cast the deciding vote. "Toronto was big," Gord said, summarizing the impact of the bylaw win in Canada's largest city as a catalyst for further action elsewhere. The other big win for Toronto was that Gord Perks is now a city councillor.

There are now over 150 municipalities in Canada with bylaws banning the cosmetic use of pesticides.[30] And a ban on the cosmetic use of pesticides effectively translates into a 2,4-D ban, since 2,4-D is the primary pesticide used for cosmetic purposes in Canada. In 2003 Quebec became the first large jurisdiction in North America to ban the cosmetic use of pesticides on public lands when the provincial government consolidated all the individual municipal bans into one province-wide ban. The legislation was extended to cover private and commercial lawns in 2006. Since 1995 the number of Quebec homeowners who admit to using pesticides on a regular basis has dropped from 30 per cent to 15 per cent.[31]

Watching 2,4-D use decline is, of course, highly troubling to the chemical industry. So disturbing, in fact, that in 2008 Dow Agrosciences, a manufacturer of 2,4-D, launched a $2 million lawsuit under the North American Free Trade Agreement (NAFTA), claiming that Quebec's provincial pesticide law was illegal.[32] Dow claimed that the Quebec ban was not based on science. This was a curious position, given that a number of groups representing or including physicians had supported bans on the cosmetic use of pesticides: the Canadian Paediatric Society, the Canadian Cancer Society, the Canadian Nurses Association, the Ontario College of Family Physicians, the Canadian Public Health Association and many other professional medical associations.[33] But for chemical companies facing pesticide regulations, when in doubt, sue somebody.

This lawsuit was launched at the same time that Ontario, the most populous province in Canada, announced the details of its province-wide pesticide ban. According to its supporters the new Ontario law would be the toughest in North America, banning approximately three hundred products, including those containing 2,4-D. When asked why 2,4-D had been included on the list of banned chemicals, Environment Minister John Gerretsen said, "We're convinced that the items that we're putting on the list are the right ingredients and right products. We think we're doing the right thing, and we won't be dissuaded by a potential NAFTA challenge. It's all about protecting kids playing in their own yards or other properties." Dow Agroscience's lawsuit appears to be an attempt to prevent similar legislation from migrating across the country and worse, for them, into the United States, where cosmetic pesticide bans are rare. In November 2008 Alberta joined Ontario and Quebec in announcing its intention to ban the use of "weed and feed" pesticide products, meaning that nearly 75 per cent of Canadians will soon be living in communities with pesticide restrictions that would have been unthinkable even a few short years ago.

The moral of this story is that if you leave setting the rules of the game to chemical manufacturers, it's almost guaranteed that the future will look a lot like the present. But happily, the game itself is finally being questioned by astute community leaders and thoughtful politicians. And citizens are the winners, hands down.

Double Exposure

You definitely do not want to spill 2,4-D on yourself. But spills do happen. Based on numerous studies on the effects of toxic substances, there are three fundamental categories of exposure: accidents; ongoing, nonaccidental exposure through use at work; and the day-to-day exposure everyone around the world faces. Accidental exposure refers to cases where people using pesticides have spilled them onto their skin or clothes. Studies and clinical assessments of people who have experienced this show that 2,4-D causes nasty immediate effects, such as headache, nausea, vomiting and eye irritation. The chemical is easily absorbed into the body through the skin and the lungs. In a number of cases, it has also disrupted an individual's ability to walk, and these debilitating effects have lasted for three years or more. Because these incidents are considered accidental when risk assessments are conducted, in the eyes of the pesticide industry, they don't count. Any reports or information prepared by industry will always say that a chemical is safe under "approved uses" or "when used according to directions." This means that a person handling the substance must wear gloves, safety goggles and full-length clothing, and accidental spills are certainly not included. For people who don't take all these precautions, exposure is more likely.

Workplace exposure refers to the regular direct exposure of farmers and pesticide applicators (including people who spray pesticides on lawns for a living). In addition to being much more likely to spill pesticides on themselves (not because they are clumsy but because they are always using them), this group is exposed to relatively high levels of 2,4-D on a regular basis. They breathe it in and get it on their clothes and possibly their skin if they are not adequately protected. Numerous studies point to a wide range of health problems for this group. The most serious is that they are two and a half times more likely to get non-Hodgkin's lymphoma, a form of blood cancer. People who work with pesticides also have more difficulty having healthy babies. In farming families in North America, for instance, there is a higher incidence of miscarriages and birth defects than in the general population. Farmers in Ontario who use pesticides also have lower sperm counts and poorer-quality sperm than non-farmers.[34]

In addition to exposure through accidents and ongoing, everyday use at work, everyone throughout most of the world experiences chronic, low-level exposure to pesticides. On a day-to-day basis, people encounter pesticide residues on food and in dust and, until recently, in some jurisdictions, pesticides sprayed on their own lawns or their neighbours' lawns. Sadly, infants are also exposed to pesticides—through their mothers' milk.[35] The potential health effects associated with chronic low-level exposure are the most difficult to identify and understand, but they include hormone-mimicking, endocrine-disruption and neurodevelopmental problems. In the case of all these disorders, chemicals confuse the body as it is developing and interfere with critical functions related to the brain, the nervous system and/or the reproductive system.

There is a clear link between pesticide exposure and many forms of cancer, and this has made doctors concerned. A recent examination by a team of doctors and scientists of more than one hundred studies on pesticides and cancer concluded that most studies on non-Hodgkin's lymphoma and leukemia showed positive associations with pesticide exposure. Children's and pregnant women's exposure to pesticides was positively associated with the cancers examined in some studies, as was parents' exposure to pesticides at work. The most consistent associations were found for brain and prostate cancer.[36]

When the team of doctors looked at research studies on non-cancer health effects of pesticides, the results were even more disturbing and the conclusions more direct. After reviewing 124 studies, doctors found the consistency of evidence showing that pesticide exposure increases the risk of neurologic, reproductive and genotoxic effects. They found strong evidence linking pesticide exposure and birth defects, fetal death and altered growth. Exposure to pesticides generally doubled the level of genetic damage as measured by chromosome aberrations in white blood cells.[37] According to the chemical industry, pesticides are safe and there is no science to suggest otherwise. According to doctors, hundreds of studies link pesticides to neurological problems, cancers, birth defects and many other disorders and diseases.

This brings two questions to mind. First, what would motivate doctors to publish a research paper showing that pesticides are a problem? Let's hear from the doctors themselves: "Family physicians need evidence-based information on the health effects of pesticides to guide their advice to patients and their involvement in community decisions to restrict use of pesticides."[38] And second, in the face of this evidence, what would motivate a corporation (Dow Agrosciences) to manufacture pesticides in the first place and to sue a jurisdiction (the province of Quebec) in opposition to a ban on the cosmetic use of pesticides?[39]

After reading dozens of studies from industry and learning about the Dow Agrosciences lawsuit, it's easy to see why people wonder who is right. Are doctors fearmongers? Are pesticides actually safe? Is it possible that chemical companies have a point and concerns are overstated? I asked Gideon Forman, Executive Director of the Canadian Association of Physicians for the

Environment (CAPE), to tell me his opinion about the role chemical companies play in the pesticides debate.

"They claimed that there's no science behind this when, in fact, there's a huge amount of science connecting pesticides to a whole range of very serious illness. They claim that the alternatives don't work when, in fact, if you look at properties that are maintained without toxic pesticides they're beautiful properties. I mean some of the most high-profile properties in the province . . . have been non-toxic for years and they're beautiful. . . . Much of it is based on the sort of thinking that was used by the tobacco lobby. Much of it is word for word taken from the tobacco lobby. The first thing they say is that there's no science connecting cigarettes to lung cancer or pesticides to cancer. And then the whole thrust of what the industry does is to place doubt in the minds of legislatures. So the industry doesn't have to prove anything. They just have to raise doubt and that was exactly their game plan with tobacco."

Forman also pointed me to another group of Canadian doctors who had reviewed studies on 2,4-D specifically. Given that so many toxic studies point to the particular sensitivity of children, it is no surprise that pediatricians have a keen interest in the subject. A study published in the journal *Pediatrics and Child Health* concludes that there is a persuasive link between 2,4-D and cancers, neurological impairment and reproductive problems.[40]

In recent years one of the significant developments contributing to the momentum behind municipal and provincial pesticide bans has been the vocal support and organizational prowess of the powerful Canadian Cancer Society, a charity juggernaut with tens of thousands of volunteers across the country. I spoke with Rowena Pinto, the Senior Director of Public Affairs for the Ontario Division of the society, about why the normally conservative organization has become so front and centre on the pesticide issue. "We reviewed the research," she said simply. "We saw a lot of potential links to cancers such as childhood brain cancer, child and adult leukemia and neuroblastoma. What made formulating a position even easier for us was that the whole idea of using pesticides for the sole intent of beautifying gardens and private lawns had no countervailing health benefit." According to Pinto, the society is concerned about many cancer risk factors but "we know that pesticides contain chemicals that are possibly carcinogenic and are one of many things that can contribute to someone developing cancer. And I'm very confident that anything we can do to reduce risk of cancer by changing the things that we can control, that aren't necessary, that have no other clear health benefit, are things we should be advocating for."

Medical researchers seem quite convinced that 2,4-D and other pesticides pose a serious human health problem. And doctors do not have a financial interest in saying pesticides are harmful. So why the controversy? Even if we accept that pesticide studies may not offer one hundred per cent conclusive proof that these substances cause particular effects (as human toxicological studies rarely do), it still seems to make sense to exercise some caution. Chemical companies tell us we need all the answers before they stop selling pesticides that have been linked to cancer in so many studies. But do we really want to take that chance for the sake of stomping out a few dandelions?

Theo's Brains
Of all the potential risks associated with pesticide use, harming the brains of children is perhaps the worst. Dr. Theo Colborn is an influential American environmental health expert, trained as a zoologist with expertise in toxicology and epidemiology. She is known for helping define the issue of endocrine disruption that was highlighted in the 1996 book she co-authored, *Our Stolen Future*. In January 2006, Dr. Colborn published a research paper questioning pesticide safety and summarizing the medical science research on the neurodevelopmental effects of exposure to pesticides.[41] She focused in particular on the development of the prenatal brain and the neonatal brain. One of the critical

issues Dr. Colborn highlights is the fact that pesticides attack neurons in the brain. This does not sound good, and it isn't. According to Dr. Colborn, "Neurons process information and are the signaling or transmitting elements in the nervous system." So any impairment of neurons will affect how our brains develop and function. As an aside, mercury—described in Chapter 5—also attacks and destroys neurons, but risk assessments for pesticides carried out by government or industry do not factor in the potentially damaging effects of exposure to mercury, or other substances for that matter, as an additional assault on the brain.

The research on the effects of pesticides on brain development and subsequent mental or physical impairment is extraordinary, both because of the effects observed after minuscule doses and because of the elaborate nature of the studies undertaken. Among other things these finding suggest that we need to know at least five things about early exposure to pesticides and the health problems that can occur when babies (and fetuses) are exposed to everyday products, including their mother's milk.

First, even tiny and very brief doses of pesticides, far below current safety levels, can damage infant brains. Second, with more than eight hundred varieties and approximately 1 billion pounds of pesticides used every year in North America, it is not surprising that pesticides are found in virtually every organ and fluid of every animal (including humans) on earth. Farm workers and their children typically have much higher levels in their bodies than the average North American. Third, neuron damage can occur when the infant brain is developing, but the neurological effects may not be seen until adolescence or adulthood. In fact, most neurological defects are not detected until later in life. Fourth, health effects linked to pesticides include everything from attention deficit hyperactivity disorder (ADHD), autism and deficits in motor performance to reproductive and hormonal defects. And finally, neurological impairments in humans are extremely difficult to test for and are therefore difficult to identify.

It is impossible to achieve the levels of proof that risk managers and government officials demand before they might deem pesticides to be unsafe—that is, unless the precautionary principle is adopted, as occurred in the Canadian Supreme Court decision regarding the Town of Hudson, Quebec. In her January 2006 paper, Dr. Colborn points in the same direction, concluding her research by saying, "An entirely new approach to determining the safety of pesticides is needed," along with "a new regulatory approach." Many people who are affected or who will potentially be affected by the damaging effects of pesticides would be the first to agree.

Rachel's Intuition

Rachel Carson's *Silent Spring* was the first major public exposé of the ecological damage caused by pesticide spraying in North America. *Silent Spring* was considered controversial at the time and some critics castigated Carson for her work, but it marked the beginning of an era in which citizens no longer accepted the received wisdom about the pervasive use of pesticides.

At the time, nearly 50 years ago, Rachel Carson was writing largely from direct field observation combined with what one might call an ecological instinct. The evidence of declining songbird populations was strong, but the direct link to pesticide spraying had not been well established. She drew on the kind of intuition that is hardwired in many humans but has been replaced by the mechanistic framework that has dominated modern times. Early humans learned to rely on clues in nature and even incorporated them directly into early warning systems. The health of other species, their calls and migration patterns, were all sources of information, and they were used to help direct human behaviour.

Modern scientists, risk managers and engineers work hard to minimize the connections between the effects of pollutants on animals and their effects on humans. Rats in labs may be getting cancer from pesticides and entire species of birds may be on the brink of collapse as a result of pollution, but humans need not be concerned, the risk managers often tell us.

In another example of counterintuitive thinking, many cities in Canada still dump untreated raw sewage into open water or have sewage systems that regularly overflow upstream into the same source that supplies municipal drinking water. Planning like this makes no sense from the point of view of human and ecological health, yet it is prevalent among many risk managers, engineers, planners and politicians. When in doubt, they seem to think, *Just add chemicals or build a bigger pipe.*

The challenge today is that unlike the obvious causes of disease and death that our ancient ancestors faced, toxic chemicals may lead to slow, insidious death from diseases such as cancer or Parkinson's that take years to manifest themselves. In other instances toxic substances may cause neurological problems that are difficult to detect. Identifying causal linkages between the exposure of a mother to a pesticide and a learning disorder in a child, for instance, is virtually impossible.

If scientists have difficulty sorting these things out, the average citizen will have even more trouble determining whether and to what extent a product is safe or harmful. Many citizens still possess an innate sense of danger, and although their observations may not be entirely accurate and are therefore not permissible in a regulatory context, this innate instinct should not be completely ignored. Rachel Carson was on to something and so was Lois Gibbs at Love Canal. So was Erin Brockovich, who won a major legal battle against Pacific Gas and Electric over chromium-contaminated water, and so is Joe Kiger in Parkersburg (Chapter 3). These are citizens observing their surroundings and sometimes their own health and sensing that something is wrong. They may not always have epidemiological data or double-blind longitudinal health studies to present, but they have eyes and common sense.

Risky Business

A well-known newspaper columnist in Canada once crowned my esteemed co-author Rick the "king of junk science." This columnist writes for a serious, albeit radically right-wing newspaper, and he is infamous for his attacks on anyone attempting to protect forests, reduce the presence of synthetic chemicals in the environment, save animals or, heaven forbid, work on what he calls the "greatest environmental conspiracy of all time," climate change. One measure of any successful environmental campaign in Canada is to be singled out by him for interfering with the hallowed marketplace. He is perhaps Canada's most prominent proponent of the so-called sound science movement: a benign-sounding but aggressive effort involving industry-funded think tanks and their public bulldogs, expounding on the virtues of everything from pesticides and tobacco to nuclear power plants.[42]

The basic strategy of the "sound science" movement is to create misleading, contrarian scientific studies and to use them to oppose environmental, health and safety regulations. Take pesticides, for example. Companies or industry-funded scientists produce irrelevant and out-of-context research, designed to undermine product bans.

In the world of industrial interests and their corporate lobbyists, "sound science" is the opposite of "junk science." The terminology here is obviously of their making and is intended to place them in an advantageous position. "Junk science," it seems, is all science that does not support the perspective that all things are safe until proven without any doubt to be harmful. So-called sound science is used as the basis to attack groups who, according to unfettered free market proponents, have silly notions such as preserving the climate for future generations, making water safe to drink or preventing the spraying of cancer-causing chemicals that make lawns look pretty. Academics whose science supports the position of environmentalists are often ironically painted as self-serving.

I am often asked why we knowingly use chemicals that cause cancer and other diseases. My response, increasingly, is "risk assessment." How is it possible that the very tool designed to prevent humans from being exposed to toxic chemicals is actually responsible for their continued use? The answer is that risk assessment helps make it easy for industry to promote so-called "sound science," which often results in requests for more and more studies, thus effectively delaying regulations that would protect the health of citizens. Risk studies assess "acceptable harm" not "unacceptable safety."

Here's one example of how risk assessment may be delaying necessary regulations. DuPont, for example, references Harvard University's Harvard Center for Risk Analysis (HCRA) in its defence of the safety of PFOA and non-stick coatings, the chemicals that may be harming the people of Parkersburg (see Chapter 3).[43] The Founding Director of this agency, Dr. John Graham, was appointed by George W. Bush in 2001 to oversee environmental, health and safety regulations in the United States as head of the Office of Information and Regulatory Affairs (OIRA). According to some observers, however, Graham (no longer head of the HCRA but still head of the OIRA) intended to render powerless the U.S. Environmental Protection Agency's ability to regulate toxic substances. Graham has "an axe to grind on cancer risk assessment," said one insider.[44] I had the opportunity to hear Graham speak at a conference organized in Washington, D.C., by the Harvard Center when Graham was still its Director. Although the conference was about the precautionary principle, it seemed apparent to me that efforts were being made to convince lawmakers in Washington that the precautionary principle was unworkable and therefore of questionable value. In his opening remarks Graham made some remarkable assertions. He talked about the potential benefits of global warming and how nitrous oxide emissions from coal plants may be good for our farms. He also suggested that too little research has been done on the potential health benefits of dioxin, one of the world's most notoriously deadly substances.

The main financial sponsors of the Harvard Center for Risk Analysis are chemical, pharmaceutical, petroleum and other large industrial companies; the U.S. government—and Health Canada. The list of contributors to the Harvard Center reads much the same as the list of the world's top chemical manufacturers.[45]

As mentioned earlier some risk managers question the assumption that health effects in lab animals such as rats and mice also have the potential to appear in humans. But for years scientists have used mice and rats for experimental purposes because the genetic variation among all animals, including humans, is remarkably small. Humans, for example, share 97 per cent of the genetic makeup of an orangutan. Rats and mice share more than 80 per cent of the genetics of a human and for decades have been part of an accepted and well-established research methodology for testing chemicals, including toxins. Rats also breed rapidly and live relatively short lives. This makes it possible to test many variations of substances and different doses on genetically similar animals. Because they age rapidly as well, the process of developing tumours or other health effects takes place within a compressed timeframe.

Major corporations go to great lengths trying to convince the public or government regulatory bodies or courts that their products are safe. When it comes to scientific research using rats or mice, one of the first lines of argument corporations use is that rats aren't humans. Their claim, ironically, is that rats are probably more sensitive to the effects of various toxic substances than humans. So unlike a toughened and self-sufficient human baby, the science-lab relatives of poor little sewer-dwelling, garbage-eating rodents are actually quite sensitive to what they put in their furry little tummies. Who'd 've thought?

It's the Dose

Here's a favourite line of risk managers: "The dose makes the poison." In other words a small amount of a substance can be perfectly harmless, or in fact necessary to life, whereas a large amount can be lethal. Salt is one example of such a substance. Humans require salt in their diets, yet a large quantity of salt over a short period of time can be lethal. Some risk managers have even used water as a demonstration of this point: no water and you die, too much and you drown.

There is some truth to this, and it certainly conforms to a basic early understanding of toxicology, whereby the goal of researchers was to identify a "threshold," or level below which a substance is deemed to be safe. According to modern environmental toxicology and endocrinology, however, in the case of some toxic substances, it has been determined that there simply is no safe limit. This means that any amount greater than zero may cause harm. Even just ten years ago this was unheard of. But mercury (see Chapter 5) is a perfect example. For years medical researchers focused on finding a "safe level" of mercury in humans. But over time, as in the case of many toxic substances, the level at which some form of human harm was detected kept getting lower. Finally, the most sophisticated mercury studies determined that the safe level is zero. That is, there is no safe level.

Risk management systems don't work well when the answer is "zero," because they're designed to find a safe level that can be measured. Zero in risk management terms is portrayed as an impossibility both to achieve and to quantify. It is for this reason that risk managers oppose environmental concepts such as "zero discharge" and "virtual elimination." They prefer the term "safe level." Furthermore, the whole concept of zero is antithetical to the risk profession because as risk managers like to say, there is no such thing as zero risk.

The study of 2,4-D provides an excellent illustration of another problem presented by risk management. When determining a "safe level," the traditional assumption (following the "dose is the poison" dogma) is that the degree of harm increases with the dose. This means that the more poison you receive, the more harm it will do. This is true for many substances, but it is not true for 2,4-D or for many other toxins. A typical dose-response curve follows a straight line. That is, the bigger the dose, the more harm it causes. But this kind of linear thinking does not apply to hormone herbicides such as 2,4-D. Hormone-disrupting chemicals follow a "U curve." A very low dose may cause significant harm, a moderate dose may cause a lesser degree of harm and a large dose again may cause significant harm. This creates the U pattern.

Unfortunately, our risk assessment and risk management systems have not advanced to the point where they can accommodate these new ways of thinking. This may be due to the financial support risk management academics and consultants receive from the chemical and petroleum industries. Or perhaps it is just a result of basic human resistance to change, even in light of new and better information. At the end of the day, risk managers cling to the concept that "the dose is the poison," making it very difficult to formulate solutions that truly protect public health or the environment.

The Foxes and the Hen House

If pesticides really are a problem, why are they still being used? It's partly because, in the United States and Canada over the past dozen years, there has been a dramatic shift from direct government regulation to a combination of industry self-regulating bodies and regulatory bodies that are closely tied to the affected corporations. In Canada, for example, the Pest Management Regulatory Agency (PMRA) has been criticized by the federal Commissioner of the Environment and Sustainable Development for being too slow and unresponsive in the regulation of pesticides.[46] The PMRA relies largely on industry-funded studies to determine pesticide safety. The Commissioner pointed out that risk assessments submitted by chemical companies lacked quality assurance and independent validation. Similar criticisms were levelled at the Canadian Food Inspection Agency (CFIA) following an outbreak of listeriosis from contaminated food products that killed over a dozen people in Canada in 2008.[47] In fact, Canada's approach to the risk management of toxic chemicals has been described as "a farce," where industry can "deny harm, remain ignorant if it can, conceal whatever it can and throw up roadblocks to controls on its activities at every turn."[48]

The problem, as I see it, is that the regulators are far too close to the industries they regulate. In other words the foxes are guarding the hen house. What's more, this is the result of a deliberate policy shift on the part of governments. It's called "client-focused" government service, and the result (and often the goal) is quite simply to make sure that government bureaucracies are aligned with business interests. Or more to the point, that government agencies make sure their "clients" (i.e., the corporations they regulate) are happy. Nothing could be further from the concept of public service. The public interest is shuffled aside in this broken model of so-called government oversight.

Thankfully, the precautionary principle is a thoughtful alternative to the rigid risk assessment approach. In a nutshell it means "better safe than sorry." The most widely agreed-upon formulation of the concept is contained in Principle 15 of the Rio Declaration on Environment and Development passed in 1992 at the United Nations Conference on Environment and Development in Rio de Janeiro, Brazil: "Where there are threats of serious or irreversible damage, lack of full scientific certainty shall not be used as a reason for postponing cost-effective measures to prevent environmental degradation." Though not always easy to implement because it disrupts the status quo, the precautionary principle is finding its way into more and more environmental statutes and regulations— including the Canadian Environmental Protection Act and the recently adopted U.S. Consumer Product Safety Commission Reform Act, which banned a variety of phthalates (see Chapter 2) already on the market until such time as the chemical industry could demonstrate their safety.

In the case of 2,4-D, it is the precautionary principle that Canadian municipalities and provinces are belatedly, though successfully, invoking in order to ditch it once and for all. Simply put, citizens in communities of all shapes and sizes right across the country overwhelmingly support pesticide bans.[49] "Better safe than sorry" makes sense to people, and loudly and clearly, they are choosing the safety of their children over weed-free lawns.

It seems we may have our priorities right after all.

EIGHT: MOTHERS KNOW BEST

[IN WHICH RICK DRINKS FROM A BABY BOTTLE]

MR. MCGUIRE: *I just want to say one word to you. Just one word.*

BENJAMIN: *Yes, sir.*

MR. MCGUIRE: *Are you listening?*

BENJAMIN: *Yes, I am.*

MR. MCGUIRE: *Plastics.*

BENJAMIN: *Exactly how do you mean?*

MR. MCGUIRE: *There's a great future in plastics. Think about it.*

—*The Graduate*, 1967

THE MOMENT I KNEW we were going to win arrived on a typically overcast Tuesday morning in November 2007.

I was standing on a stage outside the provincial legislature (called "Queen's Park") in Toronto with my colleagues from Environmental Defence. For weeks we had been advertising a "Baby Rally"—a demonstration of children—for this morning. We had urged people to come out with their kids to pressure the Ontario government to ban bisphenol A (BPA).

I was very nervous. "Don't worry! They'll come," said our talented young organizer, Cassandra Polyzou. The impressively tattooed Polyzou, one of a few dynamite 20-something activists in our office, had spent the last month laying the groundwork for today's event. She'd worked the phones. Sent out countless emails. And put up acres of posters in those nooks of the city where the sleep-deprived parents of rugrats congregate, like floppy animals relaxing at watering holes.

When I'm nervous I pace and bite my nails. I was doing a lot of both on this November morning. The "Baby Rally" was an idea that had emerged from one of our office brainstorms at Environmental Defence about how we could kick the BPA debate in Canada into high gear. We had been generating good media attention with our call for a ban on this hormone-disrupting chemical in food and beverage containers. The fact that it was the main ingredient in all the market-leading baby bottles just horrified people, and the baby bottle had quickly become a powerful icon for our campaign.

But what we needed to do in the fall of 2007 was to create a "happening" in Toronto, Canada's media capital, to really engage Ontario politicians and elicit a commitment from the Premier to take strong action. Sitting around our typically cluttered meeting room table, white board filled up with scribbled ideas, we asked ourselves what politicians really like to do. What would really pique their interest? Someone jokingly said, "Kissing babies." And the Baby Rally was born.

Polyzou was right. I shouldn't have worried. As the 10 a.m. start time drew near, the sun finally peeked through the clouds and along the half-dozen paths that lead to the front doors of the legislature, where our stage was set up, the strollers started arriving.

Nearly three hundred people with their newborns, toddlers and school-aged children streamed onto the legislature's grand front lawn. A whole fleet of those funny caterpillar-like multiple baby strollers, that look for all the world like something out of a Dr. Seuss story, arrived from nearby daycare centres. Some people spread out picnic blankets. Volunteers distributed specially made "baby pickets"—small pieces of cardboard stuck to Popsicle sticks bearing the messages "Don't Pollute Me" and "Toxic Free Ontario."

Young girl at BPA rally with her "baby" picket

The world's first (and, I believe, still the only) rally against BPA was ready to roll. I remember thinking to myself that if this many harried young parents had fought their way through city traffic to be there, public concern about BPA must truly be exploding. We were riding a wave, and it was gathering strength.

My Mother Joins In

Though perhaps not quite as rowdy as a Hannah Montana concert, the rally was pretty darn cool. By virtue of its finicky participants, it was short and sweet. It's only for so long that hundreds of children can be pacified by organic cider and cookies.

Andrea Page, the well-known founder of the exercise program "Fitmom," was the dynamic emcee and held her young son on her hip throughout. In a classic example of male clued-outedness about the world of the fairer sex, I must confess to never having heard of Page when one of our staff members first mentioned her. But my wife, and every other young mom I spoke with subsequently, sure had. Thousands of women use her videos to get back in shape after the birth of their kids, and it turned out that my family had long had one of Page's DVDs in our TV cabinet.

Page really was the perfect spokesperson for this issue, powerfully chastising industry and government for not taking the health of children seriously. As I looked around the crowd, I saw many nodding heads. Since my first meeting with her, we've done a few interviews together on the topic of BPA, and it's been a pleasure to watch her rip into the chemical industry flacks with righteous motherly indignation.

A couple of other powerful women spoke: a mom who runs a business selling "green" products for babies (who remarked on the increasing consumer demand for these items) and a representative of the Ontario Coalition for Better Child Care (who talked about why daycares had started ditching their BPA-filled baby bottles).

I was up next, and spoke as both an advocate and a parent. I told the crowd that like them, I was there for my kids. "Because I want my two boys to grow up in a world full of possibility and hope, not a world where pollution is rampant and invisible toxins threaten their health." My four-year-old son, Zack, who was with me on stage, made it onto the evening TV news by hollering "That's right" into the microphone at key moments.

Dr. Pete Myers, co-author of *Our Stolen Future*, the best-selling book on hormone-disrupting pollution and an expert on BPA, flew up from Virginia for the occasion. He started off by asking the crowd, "How many of you have a relative or friend who has been touched by breast or prostate cancer?" A sea of hands went up. "What about learning disabilities?" More hands. "How about Type 2 diabetes or infertility?" Even more. "Well, that's what we're talking about," continued Myers. "That's why we're here. The link between BPA and these diseases is strong enough scientifically to think we may be able to prevent some portion of the burden of these diseases. From everything we know about this chemical, exposure in childhood increases the likelihood of kids having health problems later in life. And the good news is that if we act on this science, we can help make people healthier."

Myers told us that no country in the world, until now, had taken strong action on BPA. The chemical companies loved to hammer home this point in order to dissuade anyone from becoming the first. "The world is watching Canada closely," Myers said. "What you are doing today matters a heckuva lot."

I looked over at my mother, who'd taken the train down from suburban Toronto to lend a hand with childcare. Over the years she'd never seemed terribly impressed with my rally going, but there she was hooting and stomping. "Have you ever been to a rally before, Mom?" I yelled. "Not really, no!" she yelled back in what I bet was an echo of the experience of many people there.

As the convoy of strollers started to disperse, it was time for a few of us to hike over to one of the many surrounding government buildings for a previously arranged meeting with Premier Dalton McGuinty and Minister of the Environment John Gerretsen.

Baby Lobbyists

Now, meeting politicians and asking for stuff is a big part of what I do for a living. Most often, such encounters follow a template. They're usually a bit formal. You present your case the best you can, and the politician in question looks at you in all sincerity and gives an oblique and puzzling answer that only partly addresses the topic at hand. You come away shaking your head.

I am delighted to report that this was not one of those meetings. In fact, the get-together with the Premier that day was like something out of a *Brady Bunch* episode, and it was definitely one of the funniest lobbying experiences I've ever had.

Things started out on an embarrassing note. As you might expect the Premier is protected by a phalanx of Ontario Provincial Police officers and highly efficient young people—the "advance" staff—whose job it is to make sure his schedule runs smoothly. One of the advance guys, in what was likely the most uncomfortable thing he did all day, asked all the moms whether they (and their babies!) could immediately turn over all their five-inch-tall Popsicle pickets.

A few of the moms blinked in surprise. The babies gurgled. The guy shuffled his feet. Just standard procedure that people meeting with the Premier can't be carrying pickets, he said. Even if the people in question are six months old. The moms took pity on him. And the babies were duly stripped of their weaponry.

Three city blocks, a few elevators and stairwells and several long corridors later, the moms and their tired, hungry infants arrived at the foodless meeting room around lunchtime. A possible meltdown loomed.

We sat at a large boardroom table, the Premier and Minister of the Environment on one side and about eight moms and their kids arrayed on the other. The media cameras, which were allowed to stay for the beginning of the meeting, contributed the occasional popping flash.

One of the moms breastfed. Some of the older kids ran laps around the table. A few diapers were changed on the floor. A little girl sat directly opposite the Premier in her mother's lap and would occasionally punctuate the conversation by locking eyes with him, leaning forward and slamming her glass baby bottle on the table. Kids were laughing. Kids were crying. It was minor mayhem.

It was hard to know where to look next. But despite the distractions, or perhaps because of them, the meeting worked. The moms, who were certainly not going to take "no" for an answer, emerged smiling. And the Premier handled it well. Though he has a reputation for being a bit of a wooden guy in front of crowds, he's very engaging one on one. He marshalled just the right combination of earnestness and baby faces that the occasion demanded.

Most important, he made a commitment that all the parents in the room wanted to hear: Ontario would move to establish Canada's first Toxic Pollution Reduction Act and would immediately seek advice from experts as to what to do about bisphenol A. No need to wait for action from the federal government, the Premier announced to the assembled moms. If the advice from the experts was to ban it, he'd ban it. It was a significant moment.

When I interviewed McGuinty a few months later and asked why he'd made the BPA commitment so quickly that day, he said that the government's best scientific advice was pointing in that direction. He went on to say, "Though in an immediate sense, a lot of this new stuff is helpful to us, makes our lives more comfortable and adds to the level of convenience associated with living in the 21st century, I've always had this sinking feeling that we haven't really fully explored the potential downsides associated with using these new materials and chemicals in consumer products." He told me how he and his wife were already trying to use more glass and fewer plastics in their kitchen at home.

The meeting with the moms made a big impression on him. Reflecting back on it he quoted author Elizabeth Stone, who wrote that having a child "is to decide forever to have your heart go walking around outside your body." "So you know," he said, "there is no more compelling desire than the desire of a parent to do right by their child. If in doubt, you want to opt for what's safe. Looking into the eyes of those moms while they were holding their kids and thinking about the hopes that I have as a parent for my own children brought home to me in a very basic kind of way the responsibility to err on the side of safety."

The rally had worked.

The Premier's announcement did indeed kick the bisphenol A debate into high gear. Queen's Park had just put Ottawa on notice that it would be a race to the finish line, with the Ontario and federal governments competing to be the first to protect Canada's kids. As we'll see the federal government was up to the challenge.

Content with a good day's work, the babies went home for their afternoon naps.

Getting to Know Them

Of all the federal political parties running for election in January 2006, Stephen Harper's Conservatives seemed the least interested in the environment. Their platform document "Stand Up for Canada" gave the environment short shrift. Harper just never talked about it unless he was scoring political points by slagging the previous government's specific failures on climate change. Yet warts and all, with the largest political firestorm in a generation— the "Sponsorship" scandal—dogging the governing Liberals, it was clear throughout much of late 2005 that the Conservatives were very likely headed for an election victory. It was time to get to know them a bit better.

To say that the environmental movement in Canada didn't have much of a relationship with the federal Conservatives at that moment would be an understatement. ("Testy would be a polite term to describe it," said Tim Powers, a Conservative commentator and Ottawa-based government relations consultant we work with.) But it hadn't always been that way. In fact, the last Progressive Conservative Prime Minister, Brian Mulroney, was awarded the honour of "Greenest Prime Minister in Canadian History" by a panel of environmental experts (of which I was one) for his work on parks, pollution and biodiversity.

When his government was defeated in 1993, the Progressive Conservative Party was shattered into three—the continuing Progressive Conservatives, the Reform and the Bloc Québécois parties. The dominant one of these—Reform—was based in oil-rich Alberta and seemed mostly interested in dissing Mulroney's legacy. Throughout much of the subsequent 13-year Liberal government, relations between the Reform Party and environmentalists remained downright frosty: environmentalists convinced that Reform was a puppet of the oil patch and Reformers grumpy that environmentalists were too close to the Liberal Party. And to be honest, there was some truth to both of these perspectives.

Things didn't really improve when the Conservative Party patched itself back together again in 2003. You could cut the mutual suspicion with a knife. And so here we were in 2005, the Tories likely about to regain power and most environmental leaders not even having so much as a Tory email address to ask them out for a coffee at Tim Hortons.

It was in this context that Environmental Defence launched the Toxic Nation campaign (see Chapter 1).

Toxic Nation

When the Conservatives won office (albeit with a tenuous minority government) in January 2006, we started meeting up a storm. We sat down with senior Environment Canada bureaucrats during the critical transition period that shapes any new government's initial priorities. We introduced ourselves to the new Minister of the Environment and to the Prime Minister's Office.

Around this same time, on the nation's airwaves and in the newspapers, Toxic Nation was getting a lot of play. There was something about it that appealed to politicians of all stripes, including Conservatives. "Toxic Nation fit with the mainstreet ethos" of the new government, Tim Powers explained. "You can see a plastic water bottle; you can't see a greenhouse gas."

Dalton McGuinty felt that Toxic Nation contributed to a changing public perception of environmental issues: "The environment is no longer an abstract, esoteric, ephemeral and romantic notion," he told me. "It's no longer just about the quality of a distant stream or the health and vigour of a remote forest. It's about the air that I'm breathing into my lungs right now. It's about the quality of the water that I give to my child at 7:30 in the evening, just before she goes to bed. It's about the container that my food has been heated up in— and that I'm about to eat from. It's become part of human health."

By demonstrating that contaminants come from many parts of our lives, Toxic Nation paved the way for the bisphenol A decision. Pollution became a live discussion again at the federal level.

In early 2006, on the same day the Conservatives took office, I had an opinion piece in the *Globe and Mail,* coauthored by Adam Daifallah, a prominent Tory activist. The column was entitled "It's in Tory Genes to Go Green" and it made this strong case: "In a minority 'pizza Parliament,' where cooperation between parties will be a prerequisite to getting anything done, there exists at least one priority common to the Conservative, Liberal, NDP and Bloc platforms. . . . The issue is pollution. An area where Prime Minister-designate Stephen Harper can not only find common ground with other parties, and make a real difference in the lives of ordinary Canadians, but one where there exists an impressive Conservative tradition just waiting to be dusted off and rehabilitated."[1]

Thankfully, at least some in the new government were listening.

Chemicals Management Plan

After the media exposure from our first Toxic Nation studies, something very interesting happened. It began slowly at first but then became a regular fixture of our workday. We started to get spontaneous phone calls in the office from people wanting to have their personal levels of pollution tested. All sorts of people, young and old. Moms and dads. Plumbers and physiotherapists. And politicians. Politicians of all stripes. All of a sudden elected officials were approaching us to donate their blood, and pee in a cup, for Toxic Nation.

We were only too happy to oblige. In June 2006 we released the test results for Rona Ambrose, then the federal Minister of the Environment; Tony Clement, federal Minister of Health; Jack Layton, Leader of the New Democratic Party; and John Godfrey, the Liberal Party Environment Critic. All of them were good sports. All of them used the opportunity of their testing to garner national headlines about how seriously they were taking the pollution problem and why they wanted to solve it.

For her troubles Ambrose was savaged in some traditionally supportive Conservative media. A "classic piece of environmental horrorism, a sort of updated take on Bram Stoker's Dracula" was how one commentator framed our campaign while assailing "Rona Brockovich" for her part in volunteering for it.[2] Apparently, in the minds of some conservatives, a comparison to the famed California pollution fighter Erin Brockovich is supposed to be unflattering. A point of view that still leaves me scratching my head, but I know that this criticism left some Tory strategists smarting.

Things got a bit rocky for us at Environmental Defence as well. Some of our environmental colleagues were downright angry with us for working with the federal Conservatives at all, for trying to nudge the government to do something useful as opposed to digging a trench and starting the aerial bombardment. "We'll cooperate with anyone in the interest of progress" was my standard response, and I remain quite proud of Environmental Defence's record of working productively with all sorts of governments.

In a particularly weird moment that I haven't told anyone about until now, an anonymous environmentalist—I've never found out who—forged a letter to me from the far-right American Enterprise Institute in Washington, D.C., congratulating Environmental Defence on its successful application for a $220,000 grant for our pollution work. There never had been such an application. The grant was fictitious, and someone had gone to a lot of trouble with

Photoshop. The letter was never sent to me. This malicious attempt to portray us as Conservative partisans was widely circulated in Ottawa, and I only received it second-hand from a puzzled friend on Parliament Hill.

Bumps in the road notwithstanding, things were now moving.

In December 2006 the federal Tories introduced the Chemicals Management Plan, a dramatic overhaul of the way in which Canada assesses and regulates two hundred high-priority chemicals used commonly in everyday life. Bisphenol A was at the top of the list.

A few months later the very first people to be tested for BPA in Canada were Premier Dalton McGuinty and the two Ontario Opposition party leaders. All three gentlemen had levels of the chemical at concentrations found in some studies to potentially affect health. "I always knew you guys were out for my blood in a figurative way—but literally?" McGuinty joked later.

John Tory, the Conservative Party Leader in Ontario, even allowed media cameras to film him as he donated blood for our tests. With a ring of six or so TV cameras surrounding him, flashes popping, reporters completely focused on the needle sticking out of his arm, I thought the poor nurse who was drawing the blood was going to pass out with the stress of it all. "I'm disappointed my blood is Liberal Red and not Conservative Blue," Tory winked at the cameras.

The Rise of BPA

Before we continue with the BPA story in Canada, let's take a step back for a second and talk about BPA 101: what it is, where it comes from and why it should concern us all.

Though the levels of this chemical in the three Ontario political leaders became the first BPA data publicly available in Canada, the Centers for Disease Control and Prevention (CDC) in the U.S. has done extensive testing. A stunning 93 per cent of Americans tested have measurable amounts of BPA in their bodies.[3] Canadians are almost certainly polluted at similar levels. The fact that virtually all of us have measurable levels of BPA all the time is even more astounding when you consider that BPA is rapidly metabolized by the human body—in just a few hours. The only possible conclusion to draw is that we are all being re-exposed on a constant basis. We're marinating in BPA every day.

Where does BPA lurk in our daily lives? In a great many places. BPA is one of the most commonly produced chemicals in the world, with industry pumping out just under 3 *billion* kilograms in 2004 versus 45 million kg in 1970—an astronomical increase in just over 30 years. In the U.S. today, it is estimated that about 70 per cent of BPA is used to manufacture polycarbonate plastic (the hard, clear plastic often marked with recycling symbol #7), about 20 per cent is used in epoxy resins and 5 per cent is used in other applications. Manufacturers of the chemical include some of the largest companies going: Bayer, Dow Chemical, General Electric, Hexion Specialty Chemicals, BASF and Sunoco Chemicals.

A typical house is chock full of BPA. Polycarbonate plastic is used to make CDs and DVDs, water bottles, drinking glasses, kitchen appliances and utensils, eyeglass lenses (like the ones I have on at the moment), bottled water carboys (the big water jugs used in office water coolers), hockey helmet visors, baby bottles, medical supplies and the faces of my laptop and Blackberry. Polycarbonate is also extensively used in cars and trucks for things like headlights, and it's right there in my kids' toy bin—in the windshields of their tiny cars, for example. Epoxy resins are used as adhesives in sporting equipment, airplanes and cars. They're also commonly found in dental filling materials, protective coatings around wire and piping and what is likely the primary avenue of exposure for most people: the interior lining of virtually every tin can found in every home and grocery store.

The explosive growth of BPA-laden products is a fairly new phenomenon. Though BPA was first synthesized way back in 1891 and its hormone-disrupting properties discovered in the 1930s, it was awhile before the true commercial worth of BPA was appreciated.

Large-scale production of epoxy resins began in the 1950s, and researchers also discovered at about this time that if you polymerize BPA (link the molecules together into long strings), you can make a hard and durable plastic—polycarbonate (polycarbonate plastic is mostly 100 per cent BPA). With increasing demand for plastics in the 1960s and 1970s, production of BPA took off. And now it's everywhere.

Now I bet at this point you're asking yourself the obvious question: What in God's name were they thinking when they started making household plastics out of a chemical that has been known for over 70 years to screw up the human body's hormone system? The short answer is that they weren't thinking about that at all. And to the extent that any brain synapses were firing, the assumption may have been made that BPA would remain bound in the plastic or come out at such low levels that it wouldn't cause any harm. Wrong, as it turns out, on both counts.

Low Dose

Anyone who works on the question of BPA very quickly comes across Dr. Fred vom Saal, a distinguished Professor at the University of Missouri. Vom Saal is an unusual academic. He's certainly the only denizen of the ivory tower I've ever met who flies his own small plane, frequently booting around the continent to various scientific conferences in his Cessna 210 Centurion. One of the first times I spoke with him, he was supposed to be on his way to Ottawa for some meetings we had set up with federal officials but was forced down by a thunderstorm in Watertown, New York. This *chutzpah* extends to his work. Unlike many university researchers, vom Saal doesn't shrink from the public eye, seeming quite content to take head on the double-barrelled invective that the chemical industry levels at him. For fully ten years, vom Saal has been at the epicentre of the BPA tornado.

In many ways the tornado was started by vom Saal himself. It was the late 1990s, and for a number of years, he had been looking at the effects of hormones on the behaviour of mice. He studied "a phenomenon that people found very interesting, that we now know occurs in human twins, and that is that babies transfer hormones to each other when they're in the uterus together." Despite the fact that the amount of hormone transferred is "just stunningly small," these exposures have specific and dramatic effects seemingly independent of the genetic makeup of the animals involved.

In one of his experiments, vom Saal observed that male mice, when exposed to tiny amounts of an estrogen hormone called estradiol, developed enlarged prostates. A very interesting result but one that was met with some serious skepticism by much of the scientific community. The reason is this: At the time, prostate cancer patients were given estrogens on the assumption that it would suppress testosterone and bring their prostate growth under control. Vom Saal's results, if correct, ran entirely counter to this approach.

Okay, he thought. *Let's really get a handle on this question.* In a series of trials, he gave progressively greater doses of estradiol and the synthetic hormone diethylstilbestrol (DES) to mice. He found the same phenomenon again: High doses blocked prostate growth and low doses dramatically stimulated prostate growth. "And everyone went, 'Oh, my God. Oops!'" explained vom Saal in summarizing the reaction to his study. His research contributed to new approaches in the use of antiestrogens to treat prostate cancer.

At the same time that vom Saal was undertaking these important experiments on estradiol and DES "low-dose" effects, he started looking at other synthetic chemicals that might have similar properties. In 1997 he published his first work on BPA. "In that study we demonstrated that a dose of BPA twenty-five thousand times lower than had ever been tested also stimulated prostate development exactly like low doses of estradiol. This had been missed in the high-dose [BPA] studies." If the reaction to his estradiol study was lively, the chemical industry's reaction to this BPA work was textbook crisis management. "They came after us like a freight train," said vom Saal.

From the industry's point of view, the stakes couldn't have been higher. If BPA was so toxic at such low levels, billions of dollars in profit were at risk. In an unusually candid comment at the time, John Waechter, a senior scientist with Dow Chemical and the Society of the Plastics Industry, admitted that should vom Saal be proven correct, the margin of safety for BPA in consumer products would be less than previously thought.[4]

According to vom Saal, the industry's efforts to get him to back off began even before his study was published. The chemical companies first became aware that something was up when he presented his results at a conference, and "the first thing they did was to send John Waechter down here to ask us, and this is a direct quote: 'Is there a mutually beneficial outcome we can arrive at where you delay the publication of your findings until the chemical industry approves them for publication?'" Vom Saal says he felt like a bribe was being offered. The industry claims that this was all a profound misunderstanding.

In any event, what is beyond dispute is that the industry has spent ten years commissioning research that might torpedo the "low-dose hypothesis." The only problem is that more and more researchers are buttressing vom Saal's findings with their own low-dose discoveries.

Accidental Discoveries

When I asked prominent geneticist Dr. Pat Hunt, a Professor at Washington State University, how she first became aware of BPA, she laughed loudly.

"We did it by pure accident," she said. "We were in the midst of some other studies. We were studying eggs from some normal mice and some mutant mice, and suddenly eggs from normal animals just went completely crazy and the data suddenly switched. . . . We went from seeing an abnormality rate of 1 to 2 per cent to 40 per cent in the 'normal' animals. So we knew something was up, and it took us weeks to figure out exactly what had happened." She and her colleagues looked at everything and finally concluded it was the animals themselves. "We started looking down in our animal facility," she explained. "We found that we had a temporary worker who came in and washed the caging materials and water bottles with the floor detergent, and this detergent has a high pH instead of a low pH [which is more corrosive]. It damaged the polycarbonate cages that our animals were living in and the water bottles they were drinking from, and they started to leach BPA." The year was 1998, and as Hunt summarized it, this serendipitous discovery meant her life "has never been exactly the same since."

Hunt and her colleagues published their study five years after the accident took place. "We didn't rush. We wanted to make certain we had the story correct because we realized that what we were coming out in the press and saying was that exposure to this chemical could cause miscarriages. This was sufficiently worrying that I wanted to make sure that we had all our t's crossed and i's dotted."

Over the past few years, Hunt has continued her BPA research and says that "everything we've done with this chemical since has only made me more concerned." Hunt's latest discovery is that BPA exposure can cause damage to multiple generations at the same time. For this experiment she exposed pregnant mice to BPA just as the ovaries in their developing female fetuses were producing a lifetime supply of eggs. When the exposed fetuses became adults, 40 per cent of their eggs were damaged. "With that one exposure," Hunt explained, "we're actually affecting three generations simultaneously."

As I interviewed people for this chapter, I realized that Pat Hunt's story is not atypical in the field of BPA research. Quite a few senior scientists who now have active BPA research programs were drawn into working on the chemical from different disciplines because of the significance they attributed to bisphenol A.

One example is Dr. Gail Prins, a Professor at the University of Illinois at Chicago, who already had a flourishing career as a prostate specialist when she began working on BPA. Her research has now provided disturbing evidence that "BPA increases the susceptibility to prostate cancer under certain conditions" and that it can achieve a permanent effect in cells at a molecular level, "reprogramming the prostate gland and affecting how it functions throughout life."

Dr. Ana Soto, a Medical Doctor at the Tufts University School of Medicine in Boston, is another prominent researcher who started working on BPA through a somewhat circuitous route. She and her colleagues became quite famous in the late 1980s when they became the first to discover an estrogen-like chemical, nonyl-phenol, leaching out of plastic. In fact, they discovered this after another laboratory "accident" like Pat Hunt's. Their new plastic test tubes started causing weird things to happen in their experiments until they isolated the estrogen-mimicking chemical in the plastic. This discovery launched her laboratory into investigations of other hormone-disrupting chemicals, including BPA, and her recent research has focused on the link between BPA and breast cancer. "What we observed in animals is that even with low doses of this chemical, we saw the development of precancerous lesions. It logically follows from this that human exposure to BPA increases the likelihood of developing breast cancer later in life."

Soto made the larger point that "all these artificial estrogens are producing in animals the effects of some of the current human epidemics. I mean breast cancer, attention deficit disorders, prostate cancer. . . . It's scary, I would say." In fact, the number of studies linking tiny amounts of BPA—amounts well within the range currently found in human bodies—to various illnesses have increased dramatically since Soto first began her BPA work.

As Fred vom Saal explained it, "We have reached a critical mass of over two hundred animal studies and over two hundred cell culture studies that take us through the exact molecular details of the response systems in human and animal cells that allow these cells to respond to staggeringly low doses of BPA. We understand now what happens at the molecular level."

Table 7.[5] Selected studies on low-dose bisphenol A exposure in animals and humans*

Dose (µg/kg/day)	Toxic Effects	Study Description	Study Year
0.025	permanent changes to genital tract	fetal exposure, osmotic pumps	2005[i]
0.025	changes in breast tissue that predispose cells to hormones and carcinogens	fetal exposure, osmotic pumps, changes noted at 6 months of age	2005[ii]
0.2	decreased activity of antioxidant enzymes and increased lipid peroxidation	oral exposure, 30-day duration	2003[iii]
2	increased prostate weight 30%	fetal exposure, oral route	1997[iv]
2	lower bodyweight, decrease of anogenital distance in both genders, signs of early puberty and longer estrus	fetal exposure, oral route	2002[v]
2.4	decline in testicular testosterone	fetal and neonatal exposure, gavage	2004[vi]
2.5	breast cells predisposed to cancer	fetal exposure, osmotic pumps	2007[vii]
10	prostate cells more sensitive to hormones and cancer	infant oral exposure, 3-day duration	2006[vii]
10	decreased maternal behaviours	fetal and neonatal exposure, oral route	2002[ix]
30	reversed the normal sex differences in brain structure and behaviour	oral during gestation and lactation	2003[x]
50	interference with brain structure and function	young adult female African green monkeys, surgical and hormone treatment	2008[xi]

50	inhibits the release and effective functioning of a key protein hormone, which increases insulin sensitivity and reduces tissue inflammation. (Suppression of this hormone could lead to insulin resistance and increased susceptibility to obesity-associated diseases—e.g., Type 2 diabetes and cardio-vascular disease).	human breast secretion, subcutaneous abdominal and adipose fat tissue samples	2008[xii]
50	EPA RfD	U.S. human exposure limit (not a result from an animal study, but a guideline set by the EPA)	1998[xiii]
100	pancreatic B cells are stimulated to release insulin, compromising the blood glucose homeostasis	oral exposure, 2-day duration	2008[xiv]

* Adapted from Environmental Working Group—Center for the Evaluation of Risks to Human Reproduction (CERHR) expert panel assessment of the utility of the study in the panel's review of BPA risk to human reproduction, 2006.

Low Dose and Paracelsus

So what is going on, and how could one chemical be responsible for so many different human ailments?

At the core of the BPA debate is the question of whether low levels of the chemical can achieve a biological effect. Many governments and the chemical industry say absolutely not. For centuries, the basic tenet of toxicology has been Paracelsus's 16th-century observation that "all things are poison and nothing is without poison; only the dose permits something not to be poisonous." This is generally shortened to "the dose makes the poison" and taken to mean that the higher the exposure to a certain chemical, the greater the impact. The chemical industry is fond of quoting Paracelsus as are many toxicologists who have never been educated to the concept of hormonally active chemicals.

Although this 16th-century logic makes intuitive sense for things like beer consumption or the amount of sugar I put in my wife's coffee every morning (which I regularly screw up), it increasingly does not make sense for hormone-mimicking chemicals like BPA. The simple reason for this is that humans (and all other animals, for that matter) have evolved over time to be sensitive to even very small amounts of hormone. It stands to reason, therefore, that our bodies will be similarly sensitive to synthetic chemicals that act like the real thing. And a huge number of our internal workings are driven by the subtle proddings from hormones. They bind to cell receptors and turn genes on and off—and so do hormone mimickers. A little hormone goes a long way.

The key to this is that hormones and compounds that behave like hormones stimulate different genes at different concentrations. And at high concentrations, they can be overtly toxic. That means that at a low dose you get one set of genes being turned on, with one or more effects, while at a high dose you get another set of genes, with effects that can be completely different. At very high doses genes get shut down because of the over-toxicity.

Pete Myers, co-author of *Our Stolen Future*, once got this concept across to me in a particularly evocative way. "Picture a drop of water, in which BPA is present in a concentration of one part per billion," he said. "Now tell me how many individual molecules of BPA would be in that water drop."

"A few thousand?" (Have I mentioned I'm a zoologist and not a chemist?)

"Nope."

"A few hundred thousand?"

"Not even close. Try 132 billion. And each one of those molecules is able to turn cell receptors on and off just like hormones do."

The implications of BPA having such major low-dose effects are profound. Government regulators, who have been focused for years on setting the so-called safe level of exposure to various chemicals have royally screwed up by overestimating the levels for hormone-disrupting chemicals. Because BPA has completely different effects at low levels than it does at high levels, there's no such thing as a "safe level."

As Fred vom Saal put it, "the traditional approach of just testing high doses of BPA completely got it wrong. This is the chemical that proves that the chemical risk assessment process is absolutely invalid for hormonally active chemicals. With Ana Soto's group at Tufts now showing effects [from BPA] at levels 2 million times lower than the lowest dose ever tested by a toxicologist, we're talking about a scale of error that's horrifying beyond belief."

In the United States the debate about low-dose effects and BPA came to a head in 2007 with the publication of two reports. In the first an advisory committee to the federal government's National Toxicology Program expressed "some concern" about the neural and behavioural impacts of fetal exposure to low doses of BPA. The report was made public under a cloud of accusation that some of the key contractors hired by the federal government to research and write it had links to the BPA industry.

The second report is the real blockbuster. The product of a remarkable U.S. National Institutes of Health–funded meeting of 38 of the world's top BPA researchers, the so-called Chapel Hill Consensus Statement is very strong in its warning: "The wide range of adverse effects of low doses of BPA in laboratory animals exposed both during development and in adulthood is a great cause for concern with regard to the potential for similar adverse effects in humans."[6] Specific human illnesses these experts believe may be linked to rising levels of BPA include increases in prostate and breast cancer; uro-genital abnormalities in male babies; a decline in semen quality in men; early onset of puberty in girls; metabolic disorders, including insulin-resistant diabetes and obesity; and neurobehavioural problems such as attention deficit hyperactivity disorder.

The Chapel Hill statement makes for chilling reading.

Milk Ducts

Now back to Canada to finish our story.

Juicy scientific and political conflict and potentially huge public health implications? Sounds newsworthy! That was what Martin Mittelstaedt, the veteran environment reporter at "Canada's national newspaper" the *Globe and Mail,* started thinking in 2005 with respect to BPA. No one has done the environment "beat" in Canada longer than the pleasantly rumpled and frighteningly well-researched Mittelstaedt. A journalist with 28 years under his belt, his interviewing often has a "Columbo" quality: It all seems innocent enough until he moves in for the kill. Since early 2006 he has written over 25 articles about BPA and has interviewed me many times. For a few brief minutes, I had the pleasure of turning the tables on him.

"Of the many environmental subjects you could focus on, why this chemical?" I asked. "Why has BPA kept your attention so consistently?"

"I thought it was one of the most worrisome, or possibly the most worrisome chemical, in widespread commerce," Mittelstaedt replied. "And those two things together made me look at it. The research on it [showed] adverse effects at some of the lowest levels ever seen in scientific research . . . test-tube results in the parts per quadrillion showing it having cellular activity and animal experiments showing it had activity in the low parts per trillion." Mittelstaedt told me that if researchers and governments found BPA problematic, it would open the door to look at all the hormonally active chemicals in a new way: To retest them and determine whether their regulatory exposure limits should be revised. "BPA is sort of like the keystone chemical in my mind," he said.

Mittelstaedt had another reason for thinking BPA would be of interest to his readers. "It was also a chemical that the public could understand because it has signature products such as plastic water bottles and tin cans. It's basically a product that everyone has in their homes and can look at. . . . You can look at a TV and wonder whether it contains Deca or not [see Chapter 4] . . . but you don't know. With BPA you can look at the product and see the number seven with the PC label on it and you know that it's in it . . . and you know that you're getting a dose."

Mittelstaedt remembers the "very disturbing" study that really piqued his journalistic interest. It was what he calls his "picture is worth a thousand words" moment. "If you had to pick one study that gave me pause, it was one that Ana Soto did in 2005. . . . She had a picture of the milk ducts in the mammary glands of mice that had been exposed *in utero* to 25 and 250 parts per trillion of BPA. The ones at 25 parts per trillion showed double the amount of milk duct growth as compared to mice that hadn't been exposed to BPA."

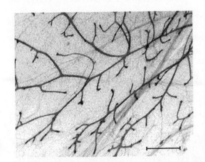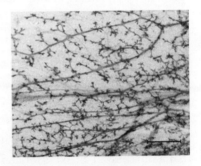

The stunning photograph that hooked Martin Mittelstaedt on BPA. Photomicrographs of whole mounts of mammary glands of mice treated with a placebo (left panel) and 25 ng BPA/kg bw·d (body weight per day) (right panel). BPA dramatically stimulates milk duct growth.[7]

If this research is substantiated, Mittelstaedt pointed out, it basically means that this chemical shouldn't be in commerce at all—because it's not possible to envision any way of protecting people when such a low active dose has such a dramatic effect.

Toxic as Tofu

Sometimes, during a campaign, manna falls from heaven. You can't plan for these surprising moments. You just need to be ready to run with them when they happen.

In June 2007 Martin Mittelstaedt broke a story that Mark Richardson (the scientist Health Canada had appointed to head up its investigation of BPA) had given a recent speech to a medical group in Tucson, Arizona, in which he endorsed the continued use of the chemical. Richardson did so using colourful language, saying that "yes, bisphenol A is estrogenic—it interacts with estrogen receptors—but a myriad of other things do as well, including proteins in tofu." He also said BPA exposures are "so low as to be totally inconsequential, in my view." Unfortunately for him, Richardson made these comments in front of a camera filming the meeting proceedings and Mittelstaedt was able to buy the DVD online for a few bucks. Pretty good investment for the *Globe and Mail*. Pretty bad day for an embarrassed federal government, which promptly yanked Richardson from the BPA file.

When Mittelstaedt first phoned me for a comment on the story, I couldn't believe what I was hearing. Not only was it exceedingly rare that a bureaucrat as experienced as Richardson would be so indiscreet; it was also a minor miracle that anyone found out about it.

We had just been handed a heaven-sent opportunity to confront the chemical industry's pro-BPA lobby machine, which had recently started to crank up in Toronto and Ottawa. I phoned all my contacts in the federal government to let them know of our concern about Richardson's views and to underline the fact that Canadians expected an objective evaluation of BPA. Fred vom Saal, Pete Myers, our Policy Director Aaron Freeman and I had a flurry of high-level meetings with political staff and senior bureaucrats in both Ottawa and Toronto over the next month to drive the point home. Judging by the grim faces of the federal bureaucrats we met, the Richardson incident was possibly the first time it had really sunk in: BPA was political dynamite and was about to shake up their world.

The Perfect Storm

By the time of the Baby Rally, as I mentioned at the beginning of this chapter, I had a feeling the tide was turning in our favour. Both the Ontario and federal governments started signalling that they were going to take action. The Chapel Hill statement was released and widely circulated, and it provided the most powerful summary to date of BPA's damaging effects.

In December Canada's largest outdoor retailer, Mountain Equipment Co-op (MEC), announced publicly that it would remove BPA-containing products from its shelves until such time as the federal government had rendered its determination of BPA's toxicity. Lululemon, the large active-wear retailer, soon followed suit. Shortly thereafter, the Environmental Working Group produced a report revealing that every major brand of infant formula tins leach BPA into the contents of the containers, and in early 2008 we released a study with U.S. colleagues measuring levels of BPA leaching out of the market-leading baby bottles.

By this point there was a prominent media story nearly every day about some aspect of BPA. Wherever you looked public confidence in BPA was quickly collapsing. A young mother I met recently told me that she was on maternity leave at about this time in Toronto and the changeover to BPA-free products by everyone she knew was "like a revolution." Seemingly overnight, stainless-steel kid's canteens and glass baby bottles became run of the mill.

Things really went nutso on April 15, 2008. This was the morning when Martin Mittelstaedt broke a story in the *Globe* confirming that the federal government would be declaring BPA a "toxic" substance within a few days and banning it in certain products, such as baby bottles. Even Mittelstaedt, who arguably knows more about BPA and its scientific and political potency than anyone else, wasn't prepared for what happened next. "I've never seen anything like it" is how he summarized the whirlwind reaction to his story that continued throughout the rest of the week.

When the government didn't deny Mittelstaedt's story, major retailers started lining up to jettison BPA products from their inventory. On the 16th it was Wal-Mart Canada, Canadian Tire, The Forzani Group (owner of several chains of sporting goods stores) and the Hudson's Bay Company.

In another dramatic development, the U.S. National Toxicology Program chose this day to release its own assessment of BPA. For the first time a U.S. government agency raised concerns regarding BPA's links to early puberty, breast cancer, prostate effects and behavioural problems and highlighted that pregnancy and early life are especially sensitive periods, given higher exposure to the chemical and limited ability to metabolize it. The U.S. media promptly started its own BPA frenzy, matching the one that was already going on north of the border.

"It's like the perfect storm," said an elated Pete Myers when I spoke with him later in the afternoon.

On the 17th, Sears Canada, Rexall Pharmacies, London Drugs and Home Depot Canada joined the parade. In our office we were completely overwhelmed with media interest, large retailers phoning to apprise us of their plans and general calls from concerned members of the public. We usually work on many issues, but that week it was all BPA all the time. Later on the 17th, I finally got the call from Ottawa I had been waiting for. "We're having a press conference to make an announcement on BPA tomorrow. We'd be glad if you could make it."

Anticipating a good announcement, I asked the Environment Minister's office if we could do anything to help them. Silence for a second. "Actually, yes" was the reply. "Could you bring some moms and babies out for the occasion? It would be great to have them there, but if we do it the national media will make a story about our cynicism."

Thinking to myself that these Conservatives were plumbing

new depths of paranoia, I replied, "Sure." And we immediately set to work, lining up some supportive moms and babies in the Ottawa area.

Better Safe than Sorry

I flew to Ottawa early that morning. By the time I landed, my cell phone voice mail was already full of media calls asking for confirmation of what the government was about to do. On the way to the press conference, I started doing radio interviews in the cab.

Aaron Freeman and I waited in the hotel lobby for our friends, the five good-natured moms who'd agreed, on short notice, to bring their babies to the press conference. A few minutes later we were all downstairs in a basement room, the moms and babies and their strollers taking up most of the front row, waiting for Health Minister Tony Clement and Environment Minister John Baird to make their entrance. Finally, the two ministers entered the room and, after the requisite amount of googoo'ing with the kids, took the podium.

As the Health Minister started to speak, I could feel the tension leaving my shoulders: "Based on the results of our assessment, today, I am proposing precautionary action to reduce exposure and increase safety. . . . We have concluded that early development is sensitive to the effects of bisphenol A. Although our science tells us exposure levels to newborns and infants are below the levels that cause effects, it is better to be safe than sorry. . . . It is our intention to ban the importation, sale and advertising of poly-carbonate baby bottles."

Success! I looked at one of the babies hanging over her mom's shoulder and winked. At that moment Canada became the first country in the world to take such action to limit exposure to bisphenol A. After years of timid indecision, Canada had finally staked out a leadership role on pollution again.

Environmental Defence's press release, sent out within minutes of the Minister's announcement, was entitled "Bisphenol A is

'Toxic.'" "Toxic" was now the legal term applied to BPA by Canada's federal pollution legislation. With such a label, I told the media in my interviews, it's only a matter of time before BPA disappears from many products. I don't care what justifications the chemical industry tries to come up with, no parent in their right mind is going to put up with their child being subjected to a "toxic" substance.

As the press conference started to break up, Aaron and I had a moment of levity when one of the more prominent national TV reporters in the room came up and asked us what we thought of the government cynically using moms and babies as a backdrop for their announcement. "The moms and babies are our peeps; they came with us," I said, grinning at Aaron, forestalling what almost certainly would have been a snarky media story. Apparently, the mutual loathing between government and the parliamentary press corps was just as out of control as the Environment Minister's office had expected.

Quite simply, the Canadian announcement was, in Fred vom Saal's words, "a bombshell." It deprived the chemical industry of their best self-fulfilling prophecy (no country has banned this substance yet, so why should anyone start now?) and reverberated around the world. As I write this chapter, BPA bans are proceeding in about a dozen U.S. states, and the Japanese government has launched a new investigation. All of these initiatives cite Canada as having shown the way. The new Obama administration is almost certain to take a harder look at BPA.

Back-of-the-Envelope Planning

As I waited for Tony Clement's announcement in that hotel basement, I was also thinking about writing this chapter. Clement's press conference took place only a few weeks after my self-experimentation with BPA. So there I was, waiting for his speech to begin, getting distracted by wondering what my BPA results were going to look like.

Our BPA experiment was, truth be told, a bit of a pioneering venture. Sure, there's been testing for BPA in the bloodstreams of people around the world. But no one has actually been dumb enough to try to seek out higher BPA levels through a variety of deliberate actions. With BPA—unlike phthalates, triclosan and mercury, where we had at least some scientific experiments to guide us—we were breaking completely new ground.

In designing the experiment I'd first called up the BPA guru himself, Dr. Fred vom Saal, to pick his brain. After laughing out loud when I told him my intentions, he started musing with me about how it could be done. I filled him in on what we already had in mind for the phthalates experiment: an initial period of "detox" to depress my phthalates levels, urine collection of this lowered level 24 hours later and then a second collection 24 hours after that, to see the effect of my phthalates exposure.

"Sounds okay," vom Saal responded. "The BPA experiment will be similar, and you can probably do it at the same time. Because the half-life of a BPA molecule in the human body is relatively short, give yourself 18 to 24 hours to try to flush it from your system. The other thing you should do to get rid of the BPA in your system is to avoid showering. It's in surface waters and you want to avoid inhaling the steam."

No shower for two days? No problem, I thought. That's more common on busy, kid-filled weekends than I care to admit.

"Then you should move into the deliberate exposure phase of the experiment. You're going to want to eat foods that are as rich in BPA as possible. Canned foods are ideal." Vom Saal told me he could help prepare a shopping list, based on the relative levels of BPA in different canned goods that he has measured. The makings for a Meal from Hell (as I came to call it).

Coffee Troubles

As I explained in Chapter 2, I decided to go two whole days eating food that had not come into contact with plastic, to try to depress levels of BPA and phthalates in my body. I won't duplicate that explanation here, but let me tell you that the cruellest blow came when the no-plastics rule disrupted my daily coffee intake. My original plan was to forego my morning coffee from coffeemakers both at home and at work. They're standard drip machines and are both made largely from plastic. Instead, what I thought I'd do was to load up on double Americanos—made fresh in a giant, expensive, stainless-steel cappuccino machine—from my favourite café on Queen Street near where I live in Toronto.

I'm in the place enough that the owner knows me, and I asked him to show me how he made my coffee—from the moment the beans came in the door of the café to the minute the cup hit my lips. I followed him around the tiny shop. First the beans arrived in bags. Then the bags are poured into the bean grinders, which look like classic grocery store bubble-gum machines—storage "tank" for the beans up top, grinder on the bottom.

Problem #1: The storage "tank," where the beans can sit for hours on end, is made of polycarbonate.

Next the beans are drained down into the grinder.

Problem #2: The receptacle that catches the crushed beans is made of polycarbonate.

From here on in, as the grounds are packed into the filter and transferred to the cappuccino machine, the beans seem to contact only metal before the beverage is poured into the paper cup. But the possibility of some serious BPA contamination was there in the grinding process.

I felt snookered. Grumpier by the second, I muddled through until the early afternoon, unclear as to how I was going to satisfy my caffeine addiction. I was saved only when our project coordinator, Sarah, came up with the idea of using a glass Bodum-style French press coffeemaker. One problem solved.

The one-litre jug of urine quickly filled up in the fridge.

Enriched with Delicious BPA

During our time in the condo test room, there's no question that Bruce ate better than I did. While he was chowing down on expensive, tasty, mercury-laden tuna steaks, I was slurping up more pedestrian fare. You can see the details in Chapter 1, but in a nutshell I ate nothing for a day and a half but canned foods heated in a polycarbonate Rubbermaid container in the microwave.[8] Campbell's chicken noodle soup, canned pineapple, Heinz spaghetti and leftover tuna casserole (not quite as good as my wife, Jen's, but not bad) cooked by Sarah with a variety of canned ingredients were the highlights. I drank a few Cokes (the cans are lined with BPA), and made my coffee in a polycarbonate French press coffeemaker purchased at Starbucks. I then drank my coffee from an old Avent polycarbonate baby bottle that Jen and I had used with our eldest son, Zack.

"Aha!" I can hear the supporters of bisphenol A crowing. "He drank his coffee out of an old baby bottle. Who does that? Smith broke his own cardinal rule of experimentation by doing something abnormal."

Not true. Most parents I know who use polycarbonate baby bottles heat them in the microwave. The hot coffee drunk from the bottle mimics the warm milk that babies receive. Also, until recently, the Starbucks near me sold a wide variety of polycarbonate travel mugs. Drinking coffee from a polycarbonate bottle is well within the bounds of normal.

"Holy Mackerel!"

So what was the outcome of this strange diet? I increased my BPA levels more than sevenfold from before exposure to after exposure. In addition to the 24-hour samples, I took three "spot" samples throughout the two-day test period. These show my BPA levels at a moment in time and, as you might expect, show a dramatic spike in BPA levels and then a decrease as my body gradually rid itself of the toxin.

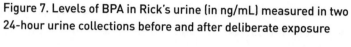

Figure 7. Levels of BPA in Rick's urine (in ng/mL) measured in two 24-hour urine collections before and after deliberate exposure

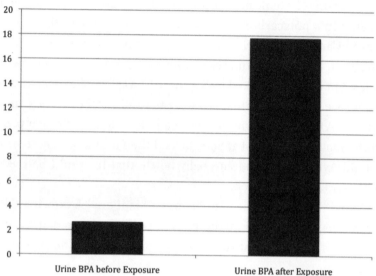

"Holy mackerel!" were Fred vom Saal's first words when I sent him the numbers. He zeroed right in on the implications for babies. "This is really scary. . . . The implications of you eating canned products and drinking out of polycarbonate the way a baby would do, and as an adult increasing the amount of BPA in your body by more than sevenfold through this procedure, are very concerning. . . . Babies are essentially doing all day, every day, what you did for one day."

Vom Saal explained that babies have a very different metabolism than adults and that the rate at which they are able to flush the BPA out of their systems and into their urine is much slower. This means that in addition to receiving high levels of BPA in 100 per cent of their food (formula from BPA-leaching cans delivered in polycarbonate bottles warmed in the microwave), any given hormone-mimicking BPA molecule is bound to stick around in a little baby body much longer than in my six-foot-six frame.

"Remarkable," said Pete Myers. "You managed to pull yourself from just below the median BPA level for the U.S. to way on top of the curve by these manipulations. But interestingly, the low levels are still reflecting some exposure and the question is: Where is that coming from?"

Figure 8. Levels of BPA in Rick's urine sampled 3 hours, 10 hours and 28 hours after initial exposure (µg BPA/g creatinine)

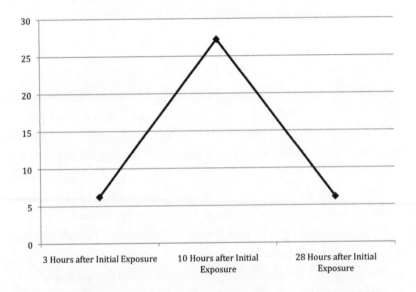

It turns out that since the last time Myers and I had chatted, some new potential sources of BPA exposure in everyday life had surfaced—sources that I hadn't been aware of during my attempted experimental "detox." So-called "carbonless" paper—the very white, glossy, coated paper that most cash register receipts are printed on these days—has very high levels of BPA. High enough levels that absorption of BPA through the skin on the fingers is likely an increasing source in daily life. Printers ink used in newspapers also contains BPA. Because these high-BPA-content papers end up in the recycling bin in many places, levels of BPA in recycled paper are generally extremely high. When I asked vom Saal about this, he agreed that contact with recycled paper could be a significant source of BPA: "When you buy a pizza, for instance, it comes in a recycled cardboard box."

I didn't eat pizza during my testing period, but I certainly handled a few newspapers and cash register receipts from the St. Lawrence Market and while I was running other weekend errands. Although I showed it's possible to reduce BPA levels in the body, it's just not possible (unfortunately!) at the moment to eliminate BPA completely and carry on a normal life without it.

Jerry Garcia Was Right

As usual, Jerry Garcia nailed it when he sang: "That's right, the women are smarter." If the beginning of the end for bisphenol A isn't a tribute to the power of concerned mothers, I don't know what is. Regardless of the excellence and persistence of Martin Mittelstaedt's reporting on the issue, the huge and escalating public concern over BPA in Canada can't be explained by stories in the traditional media alone. Within a few short weeks in the fall of 2007, it felt as if public opinion completely flipped. All of a sudden so many people had heard of bisphenol A and had an opinion. What happened?

The answer, I think, lies in the blogosphere (the vast collection of interlinked personal websites that feature regular entries of commentary) and in the rapid growth of Facebook and other social networking sites. Unlike any other issue I've worked on before, if you Google "bisphenol A," the vast majority of the bazillion hits you get are messages exchanged between individuals. People talking to one another online. From comments aimed at companies urging them to stop using BPA to messages to governments urging them to ban BPA to questions about the specific ingredients in everyday items, the blogosphere is abuzz about BPA.

"Please remove these harmful chemicals! Give all kids a better chance to grow up healthy and strong."

"You can ask your child's daycare centre to become BPA free

and to implore their suppliers to do the same. All the info plus sample letters can be found here."

"I can tomatoes from our garden, are canning lids safe? Or would they be coated as a can would?"

"We blogged it on Blog Action day. We signed the petition. We closed our eyes really tightly and tapped our shoes together three times. But it hasn't gone away yet. Bisphenol A is still all around us . . . including in baby bottles. The battle continues to be waged and we can be part of the front lines."

The mission of one of the blogs, LeagueofMaternalJustice.com (the home page of which is festooned with superheroine graphics) is "to use the power of the mom internet community to expose the injustices perpetrated against mothers everywhere and to exact vengeance through aggressive finger-wagging and online shaming."

You don't want to be running afoul of that, I tell ya.

BPA and Me

As you can probably tell, I take BPA personally. It's hard not to when I look at my two fantastic little boys and remember that Zack was raised on an Avent BPA bottle and sippy cup because Jen and I didn't know any better and Owain was not. I worry about the effects of Zack ingesting all that BPA. And the more I learn about this substance, the more my worries grow.

As my father is fond of reminding me, I'm also personally responsible for polluting *him* with BPA over the past decade. "Do you think it's responsible for my hair loss?" he once asked me with a wink. My dad is an avid canoe tripper. And sometime in the early 1990s, at Christmas I think it was, I replaced all his beat-up stainless-steel and aluminum camping plates, bowls, mugs and utensils with a matching, brand-spanking-new polycarbonate set. It seemed like the right thing to do at the time.

And that's the problem with having all these ill-understood chemicals in everyday products. You shouldn't have to be a chemical engineer to shop for your dad for Christmas or to supply your child's baby needs. What we've seen with bisphenol A in North America in the last year is many parents waking up to the fact that their governments are not doing enough to protect their children's health. And together, more quickly than anyone imagined, they are working to do something about it.

In North America, the term "soccer mom" was coined in the 1990s to describe a demographic of middle-class women who spend a significant amount of time transporting their kids to activities like soccer practice. Politicians and marketers were particularly anxious to reach them because they're an influential group that has considerable disposable income and votes in large numbers.

The significance of the BPA debate is that the soccer moms and the slightly younger parents, let's call them the "sippy cup moms," started biting back. The founder of SafeMama.com explains it this way: "I am a mother of a two-and-a-half-year-old little boy, a working mom, an author and woman of many trades. I found it completely overwhelming spending so much time researching safety issues for my child. I spent hours looking up bisphenol A or looking for the latest toy recalls. I had an 'a-ha' moment and thought that I must not be the only parent scouring the Internet for information about things that affect our children. So I started this website to keep it all in one place."[9] SafeMama.com, together with other blogs like MomsRising.org (co-founded by Joan Blades, she of MoveOn.org fame), mobilized over a hundred thousand letters to Congress in support of the *Children's Safe Products Act*, which was passed as I was writing this chapter in August 2008.

The power of moms clearly had an impact on the Canadian Conservatives. The very first thing that Conservative Environment Minister John Baird mentioned when I asked him why his government had moved against BPA was this: "I had two mothers come up to me sometime last year in a grocery store to raise this issue with me. You can see that while you have large issues like climate change, like smog, this is an in-your-face, frontline environmental concern for Canadian families."

And what companies are only beginning to understand is that this new parental community can damage or benefit brands. As one blogger put it, "the word is out—none of my friends will buy these products, nor will I. These companies risk their reputations and their profits. Mothers are networked together around the country. If they don't change—they'll see it in their bottom line."

The final word goes to Agatha Christie. Because, well, Miss Marple is a fount of knowledge about human nature: "A mother's love for her child is like nothing else in the world. It knows no law, no pity; it dares all things and crushes down remorselessly all that stands in its path."

NINE: DETOX

I was born with a plastic spoon in my mouth.

—THE WHO, 1966

IF, AT THIS POINT, you're not casting your eyes around your home, seeing things in a new light, perhaps looking at your sofa or bathroom contents with newfound suspicion, we've clearly done something wrong. Even if you were inclined to go "back to the land," as many were during the 1960s and 1970s, to try to escape "the pollution and poisoning of land, water, and air by the waste products of concentrated urban life and of large-scale industry,"[1] it wouldn't do you much good.

There's no escape from pollution. Today's most serious toxins lurk in the most private recesses of our homes. The places where we—erroneously, it turns out—feel the safest. Our exposure to these toxins is significant: The average 21st-century American can spend up to 90 per cent of his or her life indoors.[2]

Dr. Pete Myers, one of the most important figures in the modern struggle to control toxic chemicals, thinks it is exactly the demonstrable fact that we are "united in our pollution" that gives body burden testing its potency to drive the debate. "In the United States, at least," he told us, "we used to think that it was mostly poor people who were adversely impacted by toxic chemicals because of their proximity to industry, toxic waste dumps and the like. Though this remains a huge concern, we now know that everybody in America—and around the world—carries toxic chemicals inside their bodies every moment of every day. Even the wealthiest among us are affected." Myers thinks that body burden testing has illustrated in the most graphic of ways that "everyone has a problem and everyone needs a solution."

It would be easy, given the daunting nature of the toxic dilemma we've laid out, to be either paralyzed into inaction or driven to distraction with anxiety or both. But there's no need for this. We're trying to instill some concern, not worry. As we outline in this chapter, there are many things you can do to protect yourself and your family. And many that will start to take effect almost immediately.

Not infrequently during our self-experimentation and writing of this book, we've had to take our own advice in this regard. We'll admit it's difficult at times not to be overwhelmed by the enormity of the challenge created by our society's uncontrolled ocean of toxic chemicals.

As one notable example, during the days of our experiment in the condo, Rick developed what he now acknowledges to be a peculiar habit. He reserved a change of clothes for the condo, slipping into them when he arrived in the morning and out of them when he left at night. When the experiment was over, he gingerly dropped the clothes into a plastic bag, took them home and washed them at least a half-dozen times before wearing them again.

What was he thinking? He had some ill-developed notion that the heavy phthalate odours and stain-resistant coatings in the condo could be better left behind at the end of the day with a change of clothes. It was only after his wife, Jennifer, noticed him monopolizing the washing machine for hours on end and asked him what the heck he was doing that he gave his head a shake.

Nothing we did in the condo was out of the ordinary. Sure, we coated the sofa with perfluorinated chemicals and sat on it, sure Rick used lots of phthalate-containing shampoos, sure Bruce ate lots of tuna. But people do these things every day. And despite the two of us trying to avoid some of these products in our own homes, we can't avoid them when we're out and about in the world. We don't live in a bubble. Air fresheners are ubiquitous in office bathrooms, in friends' homes, in taxis. How many chairs and bus seats and carpets do we walk and sit on in the average day that are coated with stain-resistant chemicals? Lots.

The idea that we created a "chemical life" in the condo and that Rick could somehow leave it behind and return home to his "normal life" was a delusion on his part. Chemicals are everywhere.

The only unusual thing we did was to carefully monitor their rising and falling levels.

Two Conclusions

There are two basic take-home messages from our experiment.

The first is that our choices as consumers really do have a profound, and very rapid, effect on the pollution levels in our bodies. Through doing things that people do every day, Rick increased his urine levels of monoethyl phthalate (MEP) 22 times, his levels of bisphenol A 7.5 times and his levels of triclosan a mind-blowing 2,900 times. Bruce increased his mercury levels almost 2.5 times.

If we could crank up our levels of these things in a couple of days, anybody can reduce their levels—and their children's levels—of these and other chemicals in a similarly quick fashion simply by making different purchasing choices at the supermarket.

But the second conclusion flowing from our condo experiment is that no matter how hard you try, no matter how obsessively you're focused—even making the elimination of toxic chemicals from your body the single purpose of your day—you can't succeed completely. The toxins are too widespread. The sources of contamination are so numerous that no precaution taken by an individual will work completely.[3]

Though not based on any empirical evidence, in his book *Shopping Our Way to Safety*, Andrew Szasz makes a similar point: "I

am not saying that we should stop eating organic fo
willingly gorge on pesticide residues in solidarity
poisoned masses. Every person has the right to do wᵤₐ.
necessary to live their life without toxics entering their bodies. All
I am saying is that while we continue to strive to keep our and our
children's bodies healthy, we must not lose sight of the fact that
that is not enough."

Szasz views current consumer trends such as drinking bottled
water and consuming organic food as attempts by individuals to
insulate themselves from environmental problems. He wonders
to what extent, having "solved" these pollution problems for
themselves and their families, these individuals then lose interest in further collective action to substantively address the
hazards of toxins. He compares this dynamic to the debate that
erupted in the early 1960s in the United States around whether
families should build fallout shelters as a response to the dangers
of the Cold War. Most people decided this wasn't really a complete solution to the problem, and we believe this is also clear in
the case of toxic chemicals.

Making different choices the next time you go to the grocery
store can alleviate some of your family's pollution in the short
term. But for a long-term fix, only improved government regulation and oversight of toxic chemicals is the answer. It's critical that
we address this problem not only as consumers, but also as
engaged citizens demanding better of their governments.

Meanwhile—and in conclusion—here are some suggestions
for further action on the seven groups of chemicals you've just
read about.

Phthalates

The sweet smell of . . . phthalates? Phthalates aren't really what we'd
consider sweet, but they help aromas linger. Just the same, it's not
impossible to find products with sweet, natural smells that keep
your exposure to phthalates and other chemicals limited—from the

personal-care products you lather on in the mornings to the toys under foot.

Avoid personal care products with heavy artificial fragrances, especially those with "Fragrance" or "Parfum" listed as an ingredient. Manufacturers don't list phthalates among the ingredients, but these words are tip-offs to their presence. Be sure to read the label. Even if a product label says "natural" or "fragrance free," it may contain phthalates. By one estimate more than five thousand ingredients are allowed to be used in our personal-care products in North America. Forming a habit of reading labels can also help you identify other nasty ingredients that are better kept off, and out, of your body. Choose the product with the simplest ingredient list that you can.

And while you're thinking about personal-care products and bathroom rituals, how about taking a moment to *replace that polyvinyl chloride (PVC) shower curtain.* The fumes from shampoos, conditioners, soap, lotions and colognes can be potent enough without the PVC shower curtain off-gassing in that little bathroom space. Phthalates are used in PVC to soften the plastic, and the PVC odour from shower curtains is strongest after you initially open the packaging and hang it above your bathtub. While the off-gassing will dissipate over time, why not opt for something less toxic? Shower curtains can also be made from natural fibres like organic cotton and hemp or from recycled polyester.

Opt for fresh air instead of air fresheners. Those little air fresheners that get tucked into corners or plugged into electrical outlets release a lot more than simply the scent that fills the room. The Natural Resources Defense Council (NRDC) in California had 14 brands of commercially available air fresheners tested in 2007. Of these, 12 contained phthalates, including diethyl phthalate (DEP) and dibutyl phthalate (DBP) despite the absence of phthalates in the list of ingredients.[4]

Although some legislative progress has been made on keeping phthalates out of toys, we're not out of the woods yet. Canada has no regulations whatsoever, and toys containing phthalates are still permissible in the United States even after their new standards came into effect with legislation passed in the summer of 2008. The European Union, Japan, Fiji, Korea and Mexico have all implemented bans or restrictions on phthalates in children's toys and products.

Collaborating with Healthytoys.org, Mom's Rising, a U.S.-based organization, has developed *an invaluable online resource with a database that includes test results for more than 1,500 toys and products that have been tested*. You can check out specific toys and brands and also look for the best- and worst-ranked products.

Healthytoys.org includes a text-messaging service based on the Healthy Toys testing results and allows you to check out the database when you're standing in front of a product in a store. You'll find it at http://www.momsrising.org/NoToxicToys.

The United Steelworkers Union initiated a Stop Toxic Imports campaign across North America to raise awareness of lead in toys. *Hundreds of individuals have hosted "get the lead out parties" in their communities* for their colleagues, friends and family to talk about hazards in toys. Check out the campaign and get involved at www.stoptoxicimports.org.

Reduce your fat intake. Because many of the chemicals of concern travel up the food chain and are stored in fat tissues, reducing the intake of fat in foods like meat and dairy will reduce your exposure not only to phthalates but also to pesticides and PBDEs.

Action Items
- Avoid personal-care products with heavy artificial fragrance, especially those with "Fragrance" or *"Parfum"* listed as an ingredient.
- Take down that smelly PVC shower curtain and replace it with one made of recycled polyester or natural fibres.
- Unplug your air fresheners. Many air fresheners

contain phthalates. Baking soda is a natural alternative
that can be used to absorb bad odours.
- Healthytoys.org. Are you researching toys already in
your home or looking for new ones online or in a store?
Either way, you can check out the ingredients in the
products you own or the ones you might buy.
- Organize a non-toxic-toy party or a "get the lead out
party" for the other parents at your child's daycare.
- Reduce your consumption of fatty foods.
- Participate in a "rubber duck" lobby day. Join others in
your community to let your elected officials know you
want legislation that ensures children's products are
non-toxic.

The Non-stickies

Non-stick? Hardly. Perfluorochemicals (PFCs) stick around for a
very long time and they're persistent—not to mention the fact that
they're classified as a likely carcinogen.

Confused by all the perfluoro acronyms? It's easiest to avoid non-
stick altogether. And pay attention to the rapidly expanding array
of products.

Dump your old non-stick frying pan. Alternatives to toxic, non-
stick frying pans do exist. Before the miracles of science brought
us Teflon, our mothers and grandmothers fried up eggs in cast
iron or stainless steel. Cast iron can be seasoned with oil to create
a seal on which to cook, and the seal remains if no soap is used
when washing (soap breaks the seal). Flip back to Chapter 3 to
read Bruce's detailed instructions for cooking with cast iron.

Give slippery clothes the slip. Teflon has made its way into clothing.
Gore-Tex garments are made from a kind of PFC, and STAINMAS-
TER and Scotchgard are two products also made of PFCs and used
to coat or treat carpets, upholstered furniture and fabrics.

Fast food wrappers, pizza boxes and microwavable popcorn bags are also often coated with PFCs. It's the PFC coating that keeps the wrappers from getting too soaked with grease or falling apart in your hands while you're eating.

PFCs such as Teflon can be found in an increasing number of consumer items, from lipstick to windshield wiper fluid. Read the labels and avoid these products whenever you can.

There is some good news and legislative progress in the non-stick domain. In Canada PFOS (perfluorooctane sulphonate), another chemical in the non-stick family that was used in stain repellents and food packaging, was successfully added to the *Canadian Environmental Protection Act*'s Virtual Elimination List. This means the Canadian government will reduce the release of PFOS to below measurable levels.

In California a bold piece of legislation to ban PFCs from food packaging passed in the committee and the Senate only to be vetoed by Governor Schwarzenegger in late September 2008. (It just goes to show that even the greenest of governors can succumb to pressure from the chemical industry.) Had the legislation been signed, California would have been the first state to implement such a ban. We hope it won't be long before other states and Canada introduce legislation that would reduce our exposure to PFCs.

The manufacturers of PFOA in North America have agreed to a phase-out by 2015. These producers decided to regulate themselves before legislation was imposed on them. But what are they going to use as a replacement? As we were wrapping up this book, that was still unknown. But you can be sure that folks will be keeping a watchful eye on the situation. The other, less than reassuring, news is that because of the persistent and indestructible nature of PFCs, they will stick around for a long time to come and there's no assurance that the alternative chemicals DuPont and others have developed will be any safer.

Action Items

- Dump that old non-stick frying pan. Especially if it's scratched.
- Go easy on the grease. Avoid too much fast food—that hamburger, pizza or microwavable popcorn packaging may be coated with PFCs.
- Read the labels and avoid consumer products with PFCs.
- The introduction of new "replacement" chemicals provides a good opportunity to remind your politicians that chemicals should be proven safe before they're made commercially available.
- Encourage politicians to introduce legislation to phase out PFCs from food wrappers and other consumer products.

Polybrominated Diphenyl Ethers (PBDEs)

Flame retardants can be found almost everywhere in North America, but it is possible to lighten this chemical load too.

Use natural fibred products like wool, hemp and cotton. They are chemical free and naturally fire resistant. The price tag may be a little higher, but the fibres have less ecological impact. More and more companies are now manufacturing clothes, linens and household accessories made from natural fibres.

While the manufacture of some PBDEs has been phased out in North America, your old furniture or mattresses likely still contain them. *Newer furniture is more often PBDE free.* Some foams are now being manufactured without PBDEs, and an increasing number of manufacturers are selling natural foam latex made from rubber trees.

Before buying a product ask the manufacturer or retailer if it is PBDE free. Many retailers are now promoting themselves as using non toxic ingredients and materials in their products.

That great old overstuffed chair or sofa in the living room doesn't necessarily need to be replaced with wood-framed furniture. *National Geographic's Green Guide scorecard lists companies manufacturing PBDE-free furniture, electronics and other products.* It is U.S. based, and we haven't come across a scorecard for Canadian sources—but a few clicks on the Internet can turn up a number of options. Companies like IKEA, Seattle-based Greener Lifestyles and Montreal-based Essential also manufacture furniture and mattresses that are PBDE-free.

Replacing the foam or sealing or covering upholstery tears are also options to reduce exposure. Reupholstering that sofa or chair is an option too. But removing the foam from furniture can release PBDEs. So be sure that you have proper ventilation, and preferably, work in a space other than your primary living area.

Exposure to dust can also bring you into contact with PBDEs. Dust and vacuum regularly to reduce accumulation in your home. This will help reduce other toxins as well.

When you're looking to buy new electronics, ask the store or manufacturer for PBDE-free products. A number of manufacturers, including Sony, Philips, Panasonic/Matsushita and Samsung are all PBDE free. Apple is reducing PBDEs in their computers.

The Government Accountability Office in the United States estimates that 100 million TVs, computers and monitors are discarded annually.[5] Environment Canada estimates that more than 140,000 tonnes of computer equipment and other electronics end up in Canadian landfills each year.[6] There is no quick fix for this, but the problem can be mitigated by applying the principles of the "three Rs": reduce, reuse, recycle. *Reduce* the amount of PBDEs in electronic equipment or better yet eliminate them (by purchasing from PBDE-free suppliers as much as you can). *Reuse* that old computer by donating it to a local school or nonprofit organization. Or refurbish it. (There are plenty of organizations that will refurbish your old equipment and then donate it to a group in need.) *Recycle* those old computers. Where can you take this stuff instead of taking it to the dump? Many communities, local governments and an increasing number of companies have recycling programs to responsibly recycle and address "end-of-life" issues for discarding unwanted computers and other electronic equipment.

Action Items
- Use naturally fibred products—like wool, hemp and cotton. They are chemical free and naturally fire resistant.
- Buy newer, PBDE-free furniture or replace old PBDE upholstery (with proper ventilation).
- Dust and vacuum often to keep the dust and PBDEs away.
- Buy electronics that are PBDE free.
- Find a local organization that will accept and reuse your old computers and other electronic equipment.
- Write letters to politicians telling them to enact legislation to protect our homes and communities from PBDEs. There are groups all over North America working to ban PBDEs and to institute e-waste legislation.

Mercury

While mercury works its way up the food chain, a primary source of its release starts with human activities through industrial pollution. There are lots of lifestyle changes we can make to reduce our consumption of mercury and other contaminants, but until we address the issue of industrial emissions, we'll always be playing catch-up.

Eat fewer big fish and more smaller fish. Larger predatory fish contain higher levels of mercury and can also carry higher levels of other chemicals. The Environmental Protection Agency in the U.S. and Health Canada have both issued strong warnings for pregnant women to avoid mercury-laden fish. Children, particularly under the age of six, should also be given only limited amounts of tuna, choosing light over white.

The Natural Resources Defense Council in the U.S. has a handy tuna calculator that can help you determine the extent to which the amount and kinds of fish you consume are contributing to your increased body burden. You can access the calculator from its website at http://www.nrdc.org/health/effects/mercury/protect.asp.

SeaChoice has both a database and a Canadian seafood guide available on its website at www.seachoice.org that not only focuses on mercury but also looks at other chemicals of concern in fish.

The Environmental Defense Fund in the U.S. also has a comprehensive seafood selector that can be printed as a pocket guide or text-messaged to your cell phone. Check it out at www.edf.org.

If you're an avid angler, remember to check out government advisories to ensure you're catching fish deemed safe to eat. Federal, state and provincial advisories in Canada and the U.S. offer information about fish of concern, and they can tell you what sizes and locations are safe. Many large freshwater fish have elevated levels of mercury. Find a list of fish advisory resources at Environment Canada's Mercury and the Environment: Fish Consumption website (http://www.ec.gc.ca/MERCURY/EN/fc.cfm#uptomap) or the Environmental Protection Agency's Fish Advisory website (http://www.epa.gov/waterscience/fish/).

Return and recycle mercury-containing products to keep them out of the waste stream. Many manufacturers and retailers will accept old mercury-filled products such as batteries, thermostats, thermometers, compact fluorescent lights and fluorescent tubes. They do not always advertise this service but if you ask customer service at any major hardware or home improvement store, chances are they will accept and recycle your spent mercury item. If they don't, your municipality will almost certainly have a toxic waste or household hazardous waste depot where mercury and other toxic items can be returned for safe disposal. Keeping mercury out of landfills, incinerators and sewage treatment plants is critical to reducing overall mercury pollution and keeping it out of fish and our bodies.

Action Items

- Eat fewer big fish and more smaller fish. Avoid large predatory fish.
- Return used or discarded mercury-containing products to the store where you bought them or to your local household hazardous waste depot. Do not throw them in the garbage and never dump mercury in the toilet or down the sink. Once mercury goes into the garbage it will ultimately make its way back into the environment. If you dump it down the drain it will go straight into your local watershed. If you are not sure if there is mercury in the product or do not know where it was purchased, contact the manufacturer directly. When in doubt consult your local government's hazardous waste service. It is best to turn it over to them for safe disposal.
- Check out the U.S. Natural Resources Defense Council tuna calculator to see the extent to which the fish you're consuming are cranking up your mercury levels. Also take a look at the SeaChoice database and the U.S. Environmental Defense Fund seafood selector.
- Not all canned tuna is created equal. White albacore tuna should always be avoided, as it has the highest levels of mercury of any canned tuna. If you've got a tuna craving, try canned light (skipjack) tuna instead.
- Go wild. Wild fish—especially salmon—are often eco-sponsible options. Keep a seafood guide handy so you have all the information at your fingertips when you visit a fishmonger or the seafood section of your grocery store.
- Ask your grocery store/fish market to post government advisories about safe fish.

> • Mercury emissions control. Support legislators who are pushing for emissions reductions from products and industrial processes.

Triclosan

North American culture is germ obsessed. This obsession has generated a flood of products marketed as antibacterial or antimicrobial, and they aren't just lotions or potions and personal-care products. They also include items like cleaning products, and even socks, sandals and underwear. But do we really need all these antibacterial products? What happened to the good old days of warm water, soap and the 30-second scrubbing rule?

Avoid products labelled "antibacterial" that contain triclosan. Alcohol-based products aren't a problem, but those with triclosan (which will be listed as an ingredient) are. Read the label carefully, because even products not branded as antibacterial can contain triclosan, like the Gillette shave gel and Right Guard sport deodorant Rick used in his test.

Check out the Campaign for Safe Cosmetics and the Environmental Working Group's Skin Deep Cosmetic database. The Skin Deep Cosmetic database tracks ingredients, identifying toxicity and providing information from various regulatory databases for approximately twenty-nine thousand cosmetic products. Among these are not only triclosan, but also phthalates and other ingredients found in personal-care products. While it's beneficial to look for possible safety hazards in your personal-care products, it's just as good or better to search for safer products. The Campaign for Safe Cosmetics also has a safety guide for children's personal-care products: www.safecosmetics.org.

While progress can be slow, a number of companies have signed on to the *Compact for Safe Cosmetics.* By signing on, the companies have pledged not to use hazardous chemicals in their products. (This is a U.S.-based campaign, and the companies who have signed on are American, though their products are also sold in Canada.) This compact acts as an interesting barometer of how much progress is being made in cosmetics safety. Looking at the list reveals which major retailers or manufacturers have acquiesced to public pressure or have made their own ethical decisions and are signing on. *Find out if the companies whose products you purchase have signed on at* http://www.safecosmetics.org/companies/compact_with_america.cfm.

Lesstoxicguide.ca includes another great online resource listing the most common hazardous ingredients in products. The guide also identifies the worst and best Canadian products, including household cleaners and personal-care and children's products.

The practice of reading ingredients on labels should be extended to your household cleaners as well. Antibacterial hype has been applied with force to cleaning products, not just to cleaning solutions, but also to cleaning tools themselves. Triclosan is sometimes marketed under the brand name Microban. Watch for this on products like cutting boards, J Cloths, knives and even aprons.

Nanosilver and other nanoparticles have come under greater public scrutiny of late. At the same time nanosilver is being promoted by industry as an antibacterial agent and is being used in a variety of products, including socks. Some might find it amazing that such small particles could have such a powerful effect on health, but we suggest avoiding products containing nanoparticles.

Action Items

- Avoid products labelled "antibacterial" that contain triclosan, and be wary of brand names such as Microban, Biofresh, Irgasan DP 300, Lexol 300, Ster-Zac, or Cloxifermolum. Triclosan is sometimes labelled by its chemical name 5-chloro—2-(2,4-dichlorophenoxy) phenol.
- Wash your hands the "old-fashioned" way, with a good 30-second lather of soap and water.

- Check out the Campaign for Safe Cosmetics, the Skin Deep Cosmetic database and the Compact for Safe Cosmetics to find out what's in your cosmetic products.
- Check the Lesstoxicguide.ca and read labels to avoid hazardous household cleaners.
- Reach for baking soda, borax or other natural household cleaners to clean the bathroom or kitchen.
- Avoid products containing nanosilver and be wary of other nanoparticles, such as nanozinc (which is found in many sunscreens). Demand that these chemicals undergo safety testing prior to being used in products.
- Press your elected officials to legislate for better control of triclosan and nanotechnology.

Pesticides

Many individuals choose not to spray pesticides on their own lawns and gardens. This goes a long way in reducing exposure for adults, children and pets.

Go natural with chemical-free lawns. Let nature take its course, adapting your garden to your climate zone. By planting native species (plants that grow naturally in your area), you may also reduce the need for pesticides. Or choose plants that are insect resistant or that help repel pests from other plants in your garden (this is called "companion planting"). Remember that one person's weed is another's delight, so don't be shy about planting wild things that others may not enjoy as much. But note that in some jurisdictions, certain native plants are still banned from lawns and gardens.

If you must use a product in your garden or on your lawn, choose non-toxic products or the least toxic ones possible.

Municipal councils and provincial governments are implementing bans on private property and on public property such as public parks and schoolgrounds. In Canada, for instance, over 140 municipalities and 2 provinces have instituted bans on cosmetic uses of pesticides on private property within their jurisdictions. To find out more about the effects of a variety of pesticides, check the Pesticide Action Network's Pesticide Database (www.pesticideinfo.org). *Push for pesticide-free parks and schoolyards too.*

In the United States, Beyond Pesticides and the National Coalition for Pesticide-Free Lawns have been campaigning on the issue of cosmetic pesticides use. For more information (and for lawn signs and door hangers), see http://www.beyondpesticides.org/pesticide-freelawns/. In Canada, the Coalition for a Healthy Ottawa has a website with an abundance of information about pesticide action and related events not only in Ottawa but across Canada. (See www.flora.org/healthyottawa.)

Eat local and/or organic. Avoid additives, chemicals and pesticides in your food. Nature's best—organic produce—is much freer of toxins than crops grown with pesticides, herbicides and chemical fertilizers, and it usually tastes better too. And if you buy local, you are reducing the pollution created by the fossil fuel used to transport your food.

Buying organic isn't always possible when you are trying to balance your budget. *If you can't go totally organic, refer to the Environmental Working Group's Dirty Dozen list to see which vegetables and fruit tend to contain the most pesticides.* Check it out at http://www.foodnews.org/walletguide.php. Here are some samples of produce that tend to have high levels of pesticide residue: grapes (imported from outside the U.S.), peaches, strawberries, apples, spinach, nectarines, celery, pears, cherries, potatoes, sweet bell peppers and raspberries.

Thoroughly wash produce to help reduce pesticides and buy produce that's grown locally or as close to home as possible to ensure it's fresher. Many small family farms grow crops without using chemicals, although they are not certified organic.

Check out the local farmers' market to stock up on fresh produce and get connected. Lots of farmers have started to offer produce boxes that can be picked up at markets or community depots or delivered to your doorstep. If you're not sure about where or when your local farmers' market takes place, www.localharvest.org has a comprehensive website of markets, farms and community-shared agriculture in the United States. For similar information in Ontario, look at www.greenbeltfresh.ca.

Action Items

- Use an environmentally friendly lawn-care company.
- Go natural with chemical-free lawns.
- Replace your lawn with a native plant garden.
- Support local efforts to ban cosmetic use of pesticides. Email your local elected officials, requesting support for a ban on the cosmetic use of pesticides. Push for pesticide-free parks and schoolyards.
- Put a pesticide-free sign on your lawn or in your garden.
- Eat local and/or organic. Avoid pesticides and chemical additives in your food.
- Wash produce well to help remove pesticide residues.
- Clip the EWG's Dirty Dozen list and put it in your wallet, so you can avoid foods that likely contain more pesticides than others. (See http://www.foodnews.org/walletguide.php.)
- Shop at a local farmers' market and ask the vendor about pesticide use.

Bisphenol A

When buying items in plastic containers, remember this mantra: 4, 5, 1 and 2; all the rest are bad for you. Keep this mantra in mind to help you remember the recycling symbols on plastics that are especially bad for your health. Unsure of what those numbers mean? Consult the handy plastics guide.

Table 8. The handy plastics guide

Recycling Symbol	Plastic Type and Description
1 PETE	POLYETHYLENE TEREPHTHALATE Soda-pop bottles, water bottles, peanut butter jars, cooking-oil bottles, oven-ready & microwavable meal trays, detergent containers. Also used in textiles, carpet & mouldings. (A relatively safe plastic, designed for single use.)
2 HDPE	HIGH-DENSITY POLYETHYLENE Milk, juice & water jugs, detergent bottles, plastic bags, yogourt cups, shampoo bottles, cereal box liners. It's also used in piping, injection moulding, wire & cable coverings. (A relatively safe plastic, for use as food and drink containers.)
3 V	POLYVINYL CHLORIDE Water bottles, detergent & shampoo containers, cooking-oil bottles, mouthwash bottles, take-out containers, plastic wrap. It's also used in toys, piping, siding, flooring & building materials. (Avoid: May contain and/or leach a potpourri of chemicals, including bisphenol A, lead, phthalates, dioxins, mercury & cadmium. Associated with carcinogens, hormone disruptors & adverse health effects.)
4 LDPE	LOW-DENSITY POLYETHYLENE Grocery bags, container lids, plastic wrap, garbage bags, food-storage containers, coating for paper milk cartons, hot & cold drink cups, frozen-food packaging, squeezable bottles. Also used in injection moulding, wires & cable covering. (A relatively safe plastic, for use as food and drink containers.)
5 PP	POLYPROPYLENE Margarine, yogourt, syrup & other food containers, some Rubbermaid, deli & take-out containers, drinking straws, clouded plastic containers, bottle caps, medicine bottles. Also in fibres, appliances, automotive parts & carpeting. (A relatively safe plastic, for use as food and drink containers.)
6 PS	POLYSTYRENE Disposable cups, plates, bowls, cutlery, take-out containers, yogourt containers, meat trays, plastic egg cartons, foam food containers. Also used in foam packing materials, Aspirin bottles, toys, CD cases, electronic housings, insulation, coat hangers & medical products. (Avoid: Can leach styrene, a brain & nervous system toxicant, associated with adverse effects on red blood cells, liver, kidneys & stomach in animal studies. Styrene is also in second-hand smoke, car exhaust fumes, drinking water & off-gassing of building materials.)
7 PC or Other	POLYCARBONATE OR OTHER These plastics are often labelled as "other" but include polycarbonate or a combination of various resins. Three- and five-gallon water bottles, milk jugs, baby bottles, sippy cups, reusable water bottles, citrus juice bottles, the lining of tin cans, oven-baking bags. Also used in custom packaging, dental sealants, pop cans, eyewear, CDs, snowboards & car parts.

(Avoid: Polycarbonate plastic is made with bisphenol A, which can leach from the polycarbonate plastic, especially when heated. Bisphenol A is a hormone disruptor, linked to early onset of puberty, obesity, recurrent miscarriages and decreased sperm count, and it is associated with breast & prostate cancers.)

As is the case for many social justice movements, mothers have led the way in lobbying governments to regulate or ban the sale of health-threatening plastics. The chatter that used to take place around kitchen tables now happens to a greater degree on blogs, in online discussion groups and on Facebook. *There's a high volume of online discussion about bisphenol A and plastics, particularly polycarbonate plastic.*

Although it's better not to use plastic bottles at all, there's *no need to toss plastic bottles and other plastic containers into a landfill.* Our project coordinator Sarah bought an LED solar light to screw onto the opening of an old polycarbonate water bottle, and has turned the bottle into a lantern for cottaging and camping. In between it is used in the backyard. (One of the ingenious companies manufacturing solar lights—and being talked about a lot on the Internet—is www.sollight.com.)

When you're concerned about bisphenol A, consult "Z Recommends (Zrecs)" (http://zrecs.blogspot.com). Zrecs, in partnership with Mobile Commons, has created a directory of children's products, including baby bottles, that contain bisphenol A. This website is useful in Canada, as well as in the U.S. While there's a forthcoming ban on BPA baby bottles throughout Canada and some companies have pulled BPA products from their shelves, this has not been done in every store. So if you are standing in a store, trying to figure out which product is safe, use Zrecs' handy, free-of-charge, text-messaging service. (Text-messaging charges still apply, but Zrecs adds no further charges.) If you prefer not to text-message, you can download a wallet-sized, BPA-free shortlist card from the Zrecs website.

The epoxy linings of tin cans are also a source of BPA. If you use infant formula, check the U.S. Environmental Working Group's guide to infant formula and baby bottles at http://www.ewg.org/node/25724

and send an email to the companies listed there, urging them to remove BPA from their products. In Canada, the Toxic Nation Campaign has resources for parents trying to find BPA-free baby bottles at http://www.toxicnation.ca/node/161.

It's not just cans of infant formula that are of concern. *Watch out for canned foods that have high acidity, like tomatoes.*

Opt for glass over cans or fresh or frozen fruits and vegetables instead of canned foods.

Avoid putting plastic containers in the microwave. Despite what manufacturers might tell you, heating food in plastic is not safe. And note that cling wrap in microwaves is a problem because cling wrap is plastic. If you must use cling wrap, keep it out of direct contact with the food.

While we're on a plastics roll, also bear these things in mind:

Take a pass on plastic bags. Carry a reusable cloth or canvas bag with you in your purse, diaper bag or briefcase. Many communities and countries are banning or restricting plastic bags, including China, where there is an outright ban, and Ireland, where there are taxes on plastic bags. In North America, San Francisco; Oakland, California; Westport, Connecticut; and the town of Rapid Leafs in northern Manitoba have banned plastic bags altogether, while other communities are moving toward recycling programs for plastic bags. Even retailers are catching onto this one by selling reusable bags for a pittance. BYOB now has a different connotation: bring your own bag. In the past year the government-operated liquor stores in Nova Scotia and Ontario announced that they would no longer be providing plastic bags at the checkout. While not all retailers are abandoning plastic bags, others are selling them at the checkout and some, such as grocery stores in New York City, are required to collect and recycle them.

Ban the bottle—disposable water bottles, that is. Bottles containing bottled water aren't made with polycarbonate, so those bottles aren't leaching BPA. However, they're still made of plastic (PETE) designed generally for single use. Unfortunately, more than half of these water bottles are never recycled, and they end up as more junk for the landfill. London and Waterloo Region in Ontario and Maple Ridge Pitt Meadows School District in British Columbia are leading the way by banning plastic water bottles at city functions, and a number of school boards are also reviewing the sale of bottled water in schools. On college campuses debates over plastic water bottles are raging, including one at Penn State University where the debate over a ban continued as we wrapped up this book. The City of Toronto passed a ban on the sale of plastic water bottles at municipal facilities in December 2008.

Action Items
- When puzzling over the small recycling numbers on the bottom of plastic containers, remember this mantra: 4, 5, 1 and 2; all the rest are bad for you.
- Hang the Handy Plastics Guide on your fridge.
- Toss, or better yet, find an alternative use for those plastic baby bottles and use glass ones instead.
- Download a copy of the Zrecs shoppers guide for your purse or wallet or text message Zrecs to find out about the product you're looking at in the store.
- Check the Environmental Working Group guide to infant formula and baby bottles. Or look for sources of BPA-free baby bottles at http://www.toxicnation.ca/node/161.
- Organize your child's daycare to go BPA free. Encourage the daycare to sign on at the Toxic Nation website (www.toxicnation.ca).
- Eat fresh or frozen food or food stored in glass bottles instead of canned foods.
- Avoid putting plastic containers in the microwave.
- Use cloth bags instead of plastic bags for shopping.
- Contact your local representative to encourage your city to ban disposable plastic water bottles.

For even more information, first check out *www.SlowDeathBy RubberDuck.com* for updated tips and resources.

Groups to Access from the Web

Centre for Health, Environment & Justice *www.chej.org*

Environmental Defence Canada's Toxic Nation Project *www.toxicnation.ca*

Environmental Health Association of Nova Scotia's Guide to Less Toxic Products *www.lesstoxicguide.ca*

Environmental Working Group *www.ewg.org*

Greenpeace International's Eliminate Toxic Chemicals campaign *www.greenpeace.org/international/campaigns/toxics*

National Geographic's Green Guide *www.thegreenguide.com*

Natural Resources Defense Council *www.nrdc.org* and its website about everyday "Green Living": *www.simplesteps.org*

World Wildlife Fund's Detox Campaign: *www.panda.org/about_wwf/where_we_work/europe/ what_we_do/wwf_europe_environment/initiatives/ chemicals/ detox_campaign/*

Databases and Other Resources on the Web

The Centers for Disease Control and Prevention: National Report on Human Exposure to Environmental Chemicals— *http://www.cdc.gov/exposurereport*

The Collaborative on Health and the Environment: CHE Toxicant and Disease Database (approximately 180 chemicals and their health effects)—*http://database.healthandenvironment.org/*
Environmental Health News service: Keep up to date on environmental health (and more) with this daily listing of media clippings covering many issues, as well as updates on scientific research (papers and commentaries)—
http://www.environmentalhealthnews.org/
European Human Biomonitoring project:
http://www.eu-humanbiomonitoring.org/
Government of Canada's Chemicals Management Plan:
www.chemicalsubstances.gc.ca
Statistics Canada: Canadian Health Measures Survey—
http://www.statcan.gc.ca/survey-enquete/household-menages/measures-mesures/measures-mesures-eng.htm or Google Canadian Health Measures Survey
Toxipedia: free toxicology encyclopedia—*http://toxipedia.org*

Further Reading

Baker, Nena. *The Body Toxic: How the Hazardous Chemistry of Everyday Things Threatens Our Health and Well-Being.* Portland, OR: North Point Press, 2008.

Barlow, Maude, and Elizabeth May. *Frederick Street: Life and Death on Canada's Love Canal.* Toronto: Harper Collins, 2000.

Barnett, Sloan. *Green Goes with Everything: Simple Steps to a Healthier Life and a Cleaner Planet.* New York: Atria Books, 2008.

Beder, Sharon. *Global Spin: The Corporate Assault on Environmentalism.* Rev. ed. White River Junction, VT: Chelsea Green Publishing Company, 2002.

Blanc, Paul D. *How Everyday Products Make People Sick: Toxins at Home and in the Workplace.* Berkeley: University of California Press, 2007.

Bookchin, Murray. *Our Synthetic Environment.* 2nd ed. New York: Colophon, 1974.

Boyd, David. *Unnatural Law: Rethinking Canadian Environmental Law and Policy.* Vancouver: University of British Columbia Press, 2003.

Bullard, Robert. *The Quest for Environmental Justice: Human Rights and the Politics of Pollution.* San Francisco: Sierra Club Books, 2005.

Carson, Rachel. *Silent Spring.* New York: Mariner Books, 1962.

Clark, Claudia. *Radium Girls: Women and Industrial Health Reform—1910–1935.* Chapel Hill, NC: University of North Carolina Press, 1997.

Colborn, Theo, Diane Dumanoski and J.P. Myers. *Our Stolen Future.* New York: Dutton/Penguin, 1996.

Commoner, Barry. *The Closing Circle: Nature, Man and Technology.* New York: Knopf, 1971.

Davis, Devra. *When Smoke Ran Like Water: Tales of Environmental Deception and the Battle against Pollution.* New York: Basic Books, 2002.

———. *The Secret History of the War on Cancer.* New York: Basic Books, 2007.

Egan, Michael. *Barry Commoner and the Science of Survival: The Remaking of American Environmentalism.* Cambridge, MA: MIT Press, 2007.

Fagin, Dan, Marianne Lavelle and the Center for Public Integrity. *Toxic Deception: How the Chemical Industry Manipulates Science, Bends the Law and Endangers Your Health.* Monroe, ME: Common Courage Press, 1999.

Friends of the Earth U.K. *Pollution and Poverty: Breaking the Link,* report, London, January 2001.

Gibbs, Lois. *Love Canal: My Story.* Albany: State University of New York Press, 1982.

Goldwater, Leonard. *Mercury: A History of Quicksilver.* Baltimore, MD: York Press, 1972.

Houlihan, Jane, Charlotte Brody and Bryony Schwan. *Not Too Pretty: Phthalates, Beauty Products and the FDA.* Washington, D.C.: Environmental Working Group, 2002.

Howard, Ross, and Michael Perley. *Poisoned Skies: Who'll Stop Acid Rain?* Toronto: Stoddart, 1981.

Kitman, James Lincoln. "The Secret History of Lead." *The Nation,* March 20, 2000.

Levy, Stuart B. *The Antibiotic Paradox: How the Misuse of Antibiotics Destroys Their Curative Powers.* Cambridge, MA: Perseus Publishing, 2002.

Lyons, Callie. *Stain-Resistant, Nonstick, Waterproof, and Lethal: The Hidden Dangers of C8.* Westport, CT: Praeger Publishers, 2007.

Malkan, Stacy. *Not Just a Pretty Face: The Ugly Side of the Beauty Industry.* Gabriola Island, BC: New Society Publishers, 2007.

Markham, Adam. *A Brief History of Pollution.* London: St. Martin's Press, 1994.

McCormick, John. *Acid Earth: The Politics of Acid Pollution.* 3rd ed. London: Earthscan, 1997.

Meharg, Andrew. *Venomous Earth: How Arsenic Caused the World's Worst Mass Poisoning.* New York: Macmillan, 2005.

Michaels, David. *Doubt Is Their Product: How Industry's Assault on Science Threatens Your Health.* New York: Oxford University Press, 2008.

Ponting, Clive. *A Green History of the World.* London: Sinclair-Stevenson, 1991.

Rampton, Sheldon, and John Stauber. *Trust Us, We're Experts!* New York: Tarcher/Putnam, 2001.

Russell, Edmund. *War and Nature: Fighting Humans and Insects with Chemicals from World War I to Silent Spring.* New York: Cambridge University Press, 2001.

Schapiro, Mark. *Exposed: The Toxic Chemistry of Everyday Products and What's at Stake for American Power.* White River Junction, VT: Chelsea Green, 2007.

Schettler, Ted, Gina Solomon, Maria Valenti and Annette Huddle. *Generations at Risk. Reproductive Health and the Environment.* Cambridge, MA: MIT Press, 1999.

Stauber, John. *Toxic Sludge Is Good for You: Lies, Damn Lies and the Public Relations Industry.* Monroe, ME: Common Courage Press, 1995.

Steingraber, Sandra. *Having Faith: An Ecologist's Journey to Motherhood.* Cambridge, MA: Perseus, 2001.

———. *Living Downstream: A Scientist's Personal Investigation of Cancer and the Environment.* New York: Vintage, 1998.

Vasil, Adria. *Ecoholic.* Toronto: Vintage Canada, 2007.

NOTES

Preface

[1] *http://www.cancer.org/docroot/STT/STT_0.asp*

[2] Landrigan, P. J., C. B. Schechter, et al. (2002). "Environmental Pollutants and Disease in American Children: Estimates of Morbidity, Mortality, and Costs for Lead Poisoning, Asthma, Cancer, and Developmental Disabilities." Environmental Health Perspectives 110(7): 721-728.

[3] Ambachtsheer, J., J. Kron, R. Liroff, T. Little, R. Massey. March 2007. Fiduciary Guide to Toxic Chemical Risk. The Investor Environmental Health Network. *www.iehn.org*

[4] Problems with Plastic: The Controversy Over BPA Heats Up *http://www.oprah.com/article/omagazine/200904-omag-plastic-bpa*

[5] *www.cbn.com/community/closerlook/DrRoizen4/C137_ToxicFreeLiving.pdf*
[6] *http://www.epa.gov/tri/tridata/trio7/index.htm#Q1*

[7] Shapiro, Mark, Toxic inaction: Why poisonous, unregulated chemicals end up in our blood, Harper's Magazine 2007 *http://harpers.org/archive/ 2007/10/0081742*

[8] Stemming the Tide of Toxic Chemicals: A Guide to Action, Linda Greer's Blog, March 11, 2009 *http://switchboard.nrdc.org/blogs/lgreer/*

[9] *Ten Principles for Modernizing TSCA, American Chemistry Council*, August 2009. *http://www.americanchemistry.com/s_acc/sec_article_acc.asp?CID= 2178&DID=9939%20*

[10] Dow Chemical, Dow recommends principles to guide U.S. Chemicals Management Policy Reform, Friday, Aug 07, 2009

*http://www.yourindustrynews.com/dow+recommends+principles+to+guide+
u.s.+chemicals+management+policy+reform_37193.html*

[11] *EWG http://reports.ewg.org/reports/asbestos/maps/
government_data.php*

Chapter 1

[1] Theo Colborn, Dianne Dumanoski and John Peterson Myers, *Our
Stolen Future* (New York: Dutton, 1996).

[2] In 1834 the City of London brought nuisance charges against a coal gas
manufacturing operation that had contaminated the Thames River by
releasing coal tar from its factory. This case, *Rex v. Medley*, was the first
successful indictment of this type
(*http://www.radford.edu/~wkovarik/envhist/4industrial.html*).

[3] "Thames 'Clean Enough' for Salmon," *BBC News*, March 26, 2007.

[4] *http://www.great-lakes.net/teach/pollution/water/water5.html*

[5] *http://archives.cbc.ca/environment/pollution/topics/1390-8682/*

[6] *http://en.wikipedia.org/wiki/The_Lorax*

[7] David Brooks, "Clearing the Air," *New York Times*, April 20, 2004.

[8] Clive Ponting, *A Green History of the World: The Environment and the
Collapse of Great Civilizations* (London: Sinclair-Stevenson, 1991), 359.

[9] *www.en.wikipedia.org/wiki/Timeline_of_environmental_events*

[10] Elizabeth Royte, "Twenty Things You Didn't Know about Garbage,"
Discover, June 25, 2006.

[11] Lutfallah Gari, "Arabic Treatises on Environmental Pollution up to
the End of the Thirteenth Century," *Environment and History* 8, no. 2
(2002): 475–88.

[12] Ibid.

[13] Steven Johnson, *The Ghost Map: The Story of London's Most Terrifying
Epidemic—and How It Changed Science, Cities and the Modern World* (New
York: Riverhead Books, 2006).

[14] Ibid., 55.

[15] "Chemical Spill Turns Rhine Red," *On This Day*, BBC, November 1, 1986 (*http://news.bbc.co.uk/onthisday/hi/dates/stories/november/1/newsid_467900*).

[16] Benoit Nemery, Peter H.M. Hoet and Abderrahim Nemmar, "The Meuse Valley Fog of 1930: An Air Pollution Disaster," *The Lancet* 357 (March 2001): 704–8.

[17] Clarence A. Mills, "The Donora Episode," *Science* 111 (January 1950): 67–68. Mills provides a commentary on the episode a year after it occurred.

[18] *http://www.eoearth.org/article/London_smog_disaster,_England*

[19] James C. Whorton, *Before Silent Spring: Pesticides and Public Health in Pre-DDT America* (Princeton, NJ: Princeton University Press, 1975) (*http://www.radford.edu/~wkovarik/envhist/5progressive.html*).

[20] Consumers Union of United States, Inc. *Our History: 1930s.* Non-profit publisher of Consumer Reports. (*http://www.consumerreports.org/cro/aboutus/history/printable.index.htm*).

[21] Carol Ballentine, "Taste of Raspberries, Taste of Death: The 1937 Elixir Sulfanilamide Incident," *FDA Consumer Magazine* (June 1981).

[22] Peter Asch, *Consumer Safety Regulation: Putting a Price on Life and Limb* (New York: Oxford University Press, 1988), 20.

[23] Jerome Miller, "Dressed to Kill: The *Flammable Fabrics Act* of 1953—Twenty Years in Retrospect," *Cumberland Law Review* 358 (1973).

[24] Ibid.

[25] The ad ran in the Week in Review section of the *New York Times*, March 29, 1970, 6.

[26] Michael Egan, *Barry Commoner and the Science of Survival: The Remaking of American Environmentalism* (Cambridge, MA: MIT Press, 2007), 51.

[27] Ibid., 66.

[28] Andy Newman, "In Baby Teeth, a Test of Fallout: A Long-Shot Search for Nuclear Peril in Molars and Cuspids," *New York Times*, November 11, 2003.

[29] Egan, *Barry Commoner*, 67.

[30] W.K. Wyant, Jr., "50,000 Baby Teeth," *The Nation*, June 13, 1959, 535–37.

[31] From the Western Historical Manuscript Collection, University of Missouri–St. Louis.

[32] Egan, *Barry Commoner*, 72.

[33] Ibid., 71.

Chapter 2

[1] Shanna Swan, Katharina M. Main, Fan Liu, Sara Stewart et al., "Decrease in Anogenital Distance among Male Infants with Prenatal Phthalate Exposure," *Environmental Health Perspectives* 113, no. 8 (2005): 1056.

[2] "Phthalates Ester Panel: Essential 2 Know about Phthalates" (*http://www. phthalates.org/ pdfs/All_About_Phthalates.pdf*) (accessed July 30, 2008).

[3] Manori J. Silva, Dana B. Barr, John A. Reidy, Nicole A. Malek, Carolyn C. Hodge, Samuel P. Caudill, John W. Brock, Larry L. Needham and Antonia M. Calafat, "Urinary Levels of Seven Phthalate Metabolites in the U.S. Population from the National Health and Nutrition Examination Survey (NHANES) 1999–2000," *Environmental Health Perspectives* 112, no. 3 (March 2004).

[4] M. Wittassek, G.A. Wiesmüller, H.M. Koch, R. Eckard, L. Dobler, J. Müller, J. Angerer and C. Schlüter, "Internal Phthalate Exposure over the Last Two Decades: A Retrospective Human Biomonitoring Study," *International Journal of Hygiene and Environmental Health* 210, nos. 3–4 (May 2007): 319–33.

[5] L.E. Gray, Jr., C. Wolf, C. Lambright, P. Mann, M. Price, R.L. Cooper and J. Ostby, "Administration of Potentially Antiandrogenic Pesticides (Procymidone, Linuron, Iprodione, Chlozolinate, p,p´-DDE and Ketoconazole) and Toxic Substances (Dibutyl- and Diethylhexyl Phthalate, PCB 169 and Ethane Dimethane Sulphonate) during Sexual Differentiation Produces Diverse Profiles of Reproductive Malformations in the Male Rat," *Toxicology and Industrial Health* 15, nos.

1–2 (1999): 94–118; L.E. Gray, Jr., J. Ostby, E. Monosson and W.R. Kelce, "Environmental Antiandrogens: Low Doses of the Fungicide Vinclozolin Alter Sexual Differentiation of the Male Rat," *Toxicology and Industrial Health* 15, nos. 1–2 (1999): 48–64.

[6] N.E. Skakkebæk, E. Rajpert-De Meyts and K.M. Main, "Testicular Dysgenesis Syndrome: An Increasingly Common Developmental Disorder with Environmental Aspects" (Opinion), *Human Reproduction* 16, no. 5 (2001): 972–78; C.L. Acerini and I.A. Hughes, "Endocrine Disrupting Chemicals: A New and Emerging Public Health Problem?" *Archives of Disease in Childhood* 91 (2006): 633–41.

[7] Swan et al., "Decrease in Anogenital Distance among Male Infants," 1056.

[8] Swan et al., "Decrease in Anogenital Distance among Male Infants," 1056–61. Swan published another study with even stronger results in 2008: Shanna Swan, "Environmental Phthalate Exposure in Relation to Reproductive Outcomes and Other Health Endpoints in Humans," *Environmental Research* 108, no. 20 (2008): 177–84.

[9] S. Sathyanarayana, C. Karr, P. Lozano, E. Brown, A.M. Calafat, F. Liu and S.H. Swan, "Baby Care Products: Possible Sources of Infant Phthalate Exposure," *Pediatrics* 121, no. 2 (2008): e260–e268. DIBP is also fingered as a problem in another recent study: Anne-Marie Saillenfait, Jean-Philippe Sabaté and Frédéric Gallissot, "Diisobutyl Phthalate Impairs the Androgen-Dependent Reproductive Development of the Male Rat," *Reproductive Toxicology* 26, no. 2 (October 2008): 107–115.

[10] J.W. Brock, S.P. Caudill, M.J. Silva, L.L. Needham and E.D. Hilborn, "Phthalate Monoester Levels in the Urine of Young Children," *Bulletin of Environmental Contamination and Toxicology* 68 (2002): 309–14.

[11] Stephen C. Redd, "Asthma in the United States: Burden and Current Theories," *Environmental Health Perspectives* 110 (2002), Suppl. no.4: 557–60.

[12] National Institute of Environmental Health Sciences (NIEHS), "Study Shows Air Pollution Slows Lung Function Growth in Children," press release, October 20, 2000.

[13] According to Ted Schettler, the likeliest explanation of what is occurring with phthalates is that some of them interfere with testosterone synthesis. (A normal amount of fetal testosterone is essential for normal development of the male reproductive tract.)

[14] Carol Rice, Linda S. Birnbaum, James Cogliano, Kathryn Mahaffey, Larry Needham, Walter J. Rogan and Frederick S. vom Saal, "Exposure Assessment for Endocrine Disruptors: Some Consideration in the Design of Studies," *Environmental Health Perspectives* 111, no. 13 (2003): 1683–88.

[15] Giuseppe Latini, Claudio De Felice, Giuseppe Presta, Antonio Del Vecchio, Irma Paris, Fabrizio Ruggieri and Pietro Mazzeo, "In Utero Exposure to Di-(2-ethylhexyl)phthalate and Duration of Human Pregnancy," *Environmental Health Perspectives* 111, no. 14 (2003): 1783–85.

[16] National Environmental Trust, Physicians for Social Responsibility and Learning Disabilities Association of America, "Polluting Our Future: Chemical Pollution in the U.S. That Affects Child Development and Learning," September 2000 (*http://www.aaidd.org/ehi/media/ polluting_report.pdf*).

[17] Katherine M. Shea and Committee on Environmental Health, "Pediatric Exposure and Potential Toxicity of Phthalate Plasticizers," *Pediatrics* 111, no. 6 (2003): 1467–74.

[18] Toy Industry Association Inc., "Bill Banning Use of Phthalates in Toys Is Defeated in California—TIA Provides Key Testimony at January 10th Hearing," New York, January 19, 2006 (*http://www.toyassociation.org/AM/ Template.cfm?Section=Press_Releases&TEMPLATE=/CM/HTMLDisplay.cfm& CONTENTID=2468*) (accessed September 2, 2008).

[19] "Consumer Probe," *Saturday Night Live,* Episode 10, Season 2, November 12, 1976.

[20] And it's not just kids' toys I'm talking about. I had a fascinating chat with a young entrepreneur who goes by the pseudonym of Jasmine Portofino. She and a partner recently opened the first completely "environmentally friendly" sex toy shop. "We started 'Earth Erotics'

two years ago as a sort of an experiment," she told me. "It evolved out of a conversation about how unique [Portofino's hometown] Portland, Oregon, is in that we have such a green scene. We have so many green businesses. We also have a booming sex industry: boutiques, clubs, adult stores. Lots of strip clubs. So my partner and I started talking about these two huge components to our city and why it was that there aren't any environmentally friendly sex shops." Many sex toys are made with toxic ingredients. The rubberiness of dildos and vibrators results from the same phthalates as the ones used in rubber ducks, and "they're put in very absorbent and sensitive areas of the body," Portofino points out. Earth Erotics provides its customers with access to non-toxic alternatives, and after two years business "has grown exponentially."

[21] For example, the section entitled "Use of Phthalates in P&G Products" on the Procter & Gamble website (*http://www.pgperspectives.com*) states: "In contrast to DEP, laboratory studies of DBP (as well as BBP and DEHP) have shown that these compounds can cause a pattern of adverse effects in the testes and male reproductive tracts of laboratory animals."

[22] Susan M. Duty, Robin M. Ackerman, Antonia M. Calafat and Russ Hauser. "Personal Care Product Use Predicts Urinary Concentrations of Some Phthalate Monoesters," *Environmental Health Perspectives* 113, no. 11 (2005): 1530–36.

[23] One recent study concluded that the use of consumer products and different indoor sources dominates the exposure to dimethyl, diethyl, benzyl butyl, diisononyl and diisodecyl phthalates, whereas food has a major influence on the exposure to diisobutyl, dibutyl and di(2-ethylhexyl) phthalates (see Matthias Wormuth, Martin Scheringer, Meret Vollenweider and Konrad Hungerbühler, "What Are the Sources of Exposure to Eight Frequently Used Phthalic Acid Esters in Europeans?" *Risk Analysis* 26, no. 3 [2006]).

[24] Marilyn L. Wind, "Response to Petition HP 99–1: Request to Ban PVC in Toys and Other Products Intended for Children Five Years of Age and Under," August 2002, quoted in *Trouble in Toyland: NYPIRG's 2002 Toy*

Safety Report (*http://www.nypirg.org/consumer/2002/phthalates.html* [New York Public Interest Research Group (NYPIRG) website]).

[25] American Chemistry Council, "CPSC Validates Use of DINP in Vinyl Toys," press release, February 22, 2003 (*http://www.americanchemistry.com/s_phthalate/sec.asp?CID=2011&DID=8586*).

[26] Steven Milloy, "Consumer Watchdog: Vinyl Toys Are Just Ducky," February 28, 2003 (*http://www.foxnews.com/story/0,2933,79861,00.html*).

[27] I. Colón, D. Caro, C.J. Bourdony and O. Rosario, "Identification of Phthalate Esters in the Serum of Young Puerto Rican Girls with Premature Breast Development," *Environmental Health Perspectives* 108 (2000): 895–900; R. Moral, R. Wang, I.H. Russo, D.A. Mailo, C.A. Lamartinière and J. Russo, "The Plasticizer Butyl Benzyl Phthalates Induces Genomic Changes in Rat Mammary Gland after Neonatal/Prepubertal Exposure," *BMC Genomics* 8 (2007): 453.

Chapter 3

[1] "Shimmer," *Saturday Night Live,* Season 1, Episode 9, 1975.

[2] U.S. Centers for Disease Control and Prevention, *PFOA Factsheet* (*http://www.cdc.gov/exposurereport/pdf/factsheet_pfc.pdf*) (accessed December 8, 2008).

[3] Ibid.

[4] Environmental Protection Agency, "Phaseout of PFOS," correspondence from Charles Auer, May 16, 2000 (*http://www.chemicalindustryarchives.org/dirtysecrets/scotchgard/pdfs/226-0629.pdf#page=2*) (accessed December 5, 2008).

[5] Tim Lougheed, "Environmental Stain Fading Fast," *Environmental Health Perspectives* 115, no. 1 (January 2007): A20.

[6] Danish Environmental Protection Agency, *Survey of Chemical Substances in Consumer Products,* no. 99, *Survey and Environmental/Health Assessment of Fluorinated Substances in Impregnated Consumer Products and Impregnating Agents* (October 2008)

(*http://www2.mst.dk/common/Udgivramme/Frame.asp?http://www2.mst. dk/udgiv/publications/2008/978-87-7052-845-0/html/default_eng.htm*) (accessed December 15, 2008).

7 Environmental Protection Agency, "PFOA Risk Assessment" (*http://www.epa.gov/oppt/pfoa/pubs/pfoarisk.htm*) (accessed December 5, 2008). In 2005 the EPA moved the rating for PFOA up from "possible carcinogen" to "likely human carcinogen," based on its review of the evidence. DuPont disputes this categorization.

8 Anna Kärrman, Bert van Bavel, Ulf Järnberg, Lennart Hardell and Gunilla Lindström, "Perfluorinated Chemicals in Relation to Other Persistent Organic Pollutants in Human Blood," *Chemosphere* 64, no. 9 (August 2006): 1582–91.

9 "Scientists Find Rising PFC Levels in Polar Bears," *Pesticide and Toxic Chemical News*, March 31, 2008.

10 J.W. Martin, M.M. Smithwick, B.M. Braune, P.F. Hoekstra, D.C.G. Muir and S.A. Mabury, "Identification of Long Chain Perfluorinated Acids in Biota from the Canadian Arctic," *Environmental Science and Technology* 38 (2004a): 373–80.

11 Canadian Broadcasting Corporation, "PFOA—What is it and how did it get into our blood?" *Marketplace*, March 20, 2005. Interview with Erica Johnson, *Marketplace* reporter and Scott Mabury, an environmental chemist at the University of Toronto. (*http://www.cbc.ca/consumers/market/files/health/teflon/ pfoa.html*) (accessed October 20, 2008).

12 Callie Lyons, *Stain-Resistant, Nonstick, Waterproof, and Lethal: The Hidden Dangers of C8* (Westport, CT: Praeger Publishers, 2007).

13 "DuPont in Sticky Situation over Teflon Chemical" (including interview with Della Tennant), *Living on Earth*, National Public Radio, January 6, 2006.

14 Amended Class Action Complaint, Civil Action no. 01-C-2518 in the Circuit Court of Kanawha County, West Virginia, 17 (*http://www.hpcbd.com/dupont/Amended%20Complaint.PDF*).

[15] Environmental Protection Agency, "EPA Settles PFOA Case against DuPont for Largest Environmental Administrative Penalty in Agency History," news release, December 14, 2005 (*http://yosemite.epa.gov/opa/ admpress.nsf/68b5f2d54f3eefd28525701500517fbf/fdcb2f665cac66bb852570d 7005d6665!OpenDocument*).

[16] Ibid.

[17] Note that the risk assessment referred to here is the process of determining the potential health and/or environmental risks of a product or activity. These assessments are often carried out by risk consultants. If a substance is found to present a risk, risk managers are the ones who determine what actions need to be taken to reduce the problem. Risk managers may work for government, industry or academic institutions or they may be private consultants.

[18] Ken Cook, "EWG TSCA 8(e) Petition to U.S. EPA," correspondence from Ken Cook to Christine Todd Whitman, Administrator, EPA, April 11, 2003 (*http://www.ewg.org/node/21317*) (accessed December 5, 2008). The original 1961 DuPont memo is posted at *http://www.defendingscience.org/ case_studies/upload/1961-memo.pdf* (accessed December 5, 2008).

[19] "3M and Scotchgard: 'Heroes of Chemistry' or a 20-year coverup?" Chemical Industry Archives (*http://www.chemicalindustryarchives.org/dirtysecrets/scotchgard/1.asp*) (accessed August 10, 2008).

[20] Ibid.

[21] DuPont, "PFOA" (*http://www2.dupont.com/Sustainability/en_US/ positions_issues/pfoa.html*) (accessed December 8, 2008).

[22] Environmental Working Group, *PFCs: Global Contaminants: DuPont's Spin about PFOA*, research report, Washington, D.C., April 3, 2003. The EWG report describes an unpublished report in which 3M discovered these birth defects when conducting studies on rats in 1983. The results were made available by DuPont as part of the class action suit.

[23] Law Firm of Hill, Peterson, Carper, Bee & Deitzler, "C-8 Class Action Settlement"

(*http://www.hpcbd.com/C-8%20Class%20Action%20Settlement.htm*) (accessed December 5, 2008).

24 Harold Brubaker, "DuPont Settles Pollution Lawsuit: The Firm Will Pay $108 million to Resolve Allegations That Discharge from a W. Va. Plant Contaminated Water Supplies," *Philadelphia Inquirer,* September 9, 2004.

25 DuPont, *2007 Annual Review* (*http://library.corporate-ir.net/library/73/733/73320/items/283770/DD_2007_AR_v2.pdf*) (accessed November 8, 2008).

26 K.S. Betts, "Perfluoroalkyl Acids: What Is the Evidence Telling Us?" *Environmental Health Perspectives* 115, no. 5 (May 2007): A250–A256; K.S. Betts, "PFOS and PFOA in Humans: New Study Links Prenatal Exposure to Lower Birth Weight," *Environmental Health Perspectives* 115, no. 11 (November 2007): A550.

27 "First C8 Study Results Listed," *Parkersburg News and Sentinel,* October 16, 2008.

28 R.E. Wells, "Fatal Toxicosis in Pet Birds Caused by Overheated Cooking Pan Lined with Polytetrafluoroethylene," *Journal of the American Veterinary Medical Association* 182 (1983): 1248–50.

29 "Learn More about DuPont™ Teflon®" (*http://www.teflon.com/NASApp/Teflon/TeflonPageServlet?pageId=/consumer/na/eng/housewares/keyword/teflon_keyword_birds.html*).

30 Jane Houlihan, Kris Thayer and Jennifer Klein, *Canaries in the Kitchen: Teflon Toxicosis,* research report (Washington, D.C.: Environmental Working Group, 2003).

31 Ibid.

32 Houlihan, Thayer and Klein, *Canaries in the Kitchen.*

33 Ibid.

34 R.F. Brown and P. Rice, "Electron Microscopy of Rat Lung Following a Single Acute Exposure to Perfluoroisobutylene (PFIB): A Sequential Study of the First 24 Hours Following Exposure," *International Journal of Experimental Pathology* 72 (1991): 437–50.

[35] Masami Son et al., "A Case of Polymer Fume Fever with Interstitial Pneumonia Caused by Inhalation of Polytetrafluoroethylene (Teflon)," *Japanese Journal of Toxicology* 19, no.3 (2006).

[36] DuPont, "Innovative Short Chain Products" (*www2.dupont.com/Capstone/en_US/innovative_short_chain_products.html*).

[37] Antonia M. Calafat, Lee-Yang Wong, Zsuzsanna Kuklenyik, John A. Reidy and Larry L. Needham, "Polyfluoroalkyl Chemicals in the U.S. Population: Data from the National Health and Nutrition Examination Survey (NHANES) 2003–2004 and Comparisons with NHANES 1999–2000," *Environmental Health Perspectives* 115, no. 11 (November 2007): 1596–1602.

[38] M.J.A. Dinglasan-Panlilio and S.A. Mabury, "Significant Residual Fluorinated Alcohols Present in Various Fluorinated Materials," *Environmental Science and Technology* 40 (2007): 1447–53.

[39] 1. Total condo room volume was about 26.6 m^3.

2. Total surface area of material to protect was ~19 m^2 (drapes = 1 m^2, couch = 3 m^2 and carpet = 15 m^2).

3. Teflon Advanced jug recommends an application rate of 200 square feet per gallon of diluted product (1 part pure product, 4 parts water). In metric measurement this equates to ~19 m^2 per 4 L (litres) of "diluted product." So, considering the recommended application rate, 4 L of "diluted product," which is equal to 800 mL of pure product, would be applied (since the area to be covered was about 19 square metres).

4. In 800 mL of pure product there would be about 0.150 grams of fluorotelomer alcohols (FTOHs), the precursors to PFOA. Assuming this was all released (probably not in reality), the result would be an air concentration of 2.5 micrograms per cubic metre ($\mu g/m^3$). This is on the high side of what has been calculated for indoor air, Butt told us, but not absurdly high.

5. The conversion of expected indoor air concentration to a response in the blood levels is where things got tricky. Over the 24 hours of exposure

(2 days, 12 hours per day) Butt assumed that Rick would breathe about 9 m^3 of air, which at a rate of 2.5 $\mu g/m^3$, would result in exposure to 22.5 μg of FTOH. If we assume that all of the FTOH is taken up across the lungs (usually the uptake efficiency is much lower, but a worst-case scenario was being assumed here) and if we assume that all of the FTOH is converted to PFOA and other perfluorinated acids (again, this conversion efficiency is far greater than would be expected in reality), Rick would accumulate 22.5 μg of FTOH in his blood. Based on "average" human blood volumes, Butt assumed that Rick (who, at six-foot-six, is taller than most) had about 8 L of blood. So if all the perfluorinated acids accumulated in his blood (they would presumably also accumulate in the liver and kidney, lowering the actual concentrations), the concentration would increase to ~3 ng/mL, similar to Rick's existing PFOA value.

Chapter 4

[1] A.R. Horrocks and D. Price, *Fire Retardant Materials* (Cambridge: Woodhead Publishing, 2001).

[2] Chlorinated flame retardants work in a very similar manner and present similar environmental concerns. There is a great deal of concern over the use of chlorinated tris as a flame retardant.

[3] Mehran Alaee, Pedro Arias, Andreas Sjödin and Åke Bergman, "An Overview of Commercially Used Brominated Flame Retardants, Their Applications, Their Use Patterns in Different Countries/Regions and Possible Modes of Release," *Environment International* 29, no. 6 (September 2008): 683–89.

[4] "Great Lakes Chemical: An Aggressive Growth Plan Pays Off," *Chemical Week*, March 21, 1984.

[5] Linda Charlton, "Intentions Gone Astray: The Facts about Tris Don't Leave Much Choice," The Week in Review, *New York Times*, July 3, 1997, 97.

[6] Nadine Brozan, "Flame Retardant Sleepwear: Is There a Risk of Cancer?" Travel Section—Accent on the Sun, *New York Times*, April 10, 1976, 38.

[7] Ibid.

[8] "Ban Asked on Children's Wear with Flame Retardant," *New York Times*, February 9, 1977, 21.

[9] Arlene Blum, Marian Deborah Gold, Bruce N. Ames, Christine Kenyon, Frank R. Jones, Eva A. Hett, Ralph C. Dougherty, Evan C. Horning, Ismet Dzidic, David I. Carroll, Richard N. Stillwell and Jean-Paul Thenot, "Children Absorb Tris-BP Flame Retardant from Sleepwear: Urine Contains the Mutagenic Metabolite, 2,3-Dibromopropanol," *Science* 201, no. 4360 (September 15, 1978): 1020–23.

[10] Mark Hosenball, "Karl Marx and the Pajama Game," *Mother Jones*, November/December 1979.

[11] Ibid.

[12] Luther J. Carter, "Michigan's PBB Incident—Chemical Mix-Up Leads to Disaster," *Science* 92, no. 4236, April 16, 1976, 240–43.

[13] Michael R. Reich, "Environmental Politics and Science: The Case of PBB Contamination in Michigan," *American Journal of Public Health* 73, no. 3 (March 1983): 301–13.

[14] *http://www.michigan.gov/documents/mdch_PBB_FAQ_92051_7.pdf*

[15] Reich, "Environmental Politics and Science," 301–13.

[16] Jane Brody, "Perils in a Chemical World: PBB Incident in Michigan Is Viewed as Latest Evidence of Need for New Investigative System," *New York Times*, November 11, 1976, 61.

[17] Joan Lowy, "Safety of a New Flame Retardant Questioned," Scripps Howard News Service, October 11, 2004.

[18] Takesumi Yoshimura, "Yusho in Japan," *Industrial Health* 41, no. 3 (2003): 139–48.

[19] Kim Hooper and Thomas A. McDonald, "The PBDEs: An Emerging Environmental Challenge and Another Reason for Breast-Milk Monitoring Programs," *Environmental Health Perspectives* 108, no. 5 (May 2000): 387–92.

[20] Ibid.

21 The ability of PCBs and PBDEs to act as endocrine disruptors has received more public scrutiny than their neurotoxicity during development. The Yu-Cheng children were mentally handicapped, and several high-quality epidemiological studies have documented adverse effects of PCBs on IQ and other behaviours at environmental levels. The effects of PCBs and PBDEs on behaviour in animals are pretty much the same. Moreover, the biochemical effects in brain tissue are the same for both kinds of chemicals. So although the epidemiological studies on PBDEs have not been done, it is highly likely that PDBEs are developmental neurotoxicants in humans. The mechanisms are largely independent of endocrine disruption, although some behavioural effects could be produced by effects on the endocrine system.

22 Arlene Blum and Bruce N. Ames, "Flame Retardant Additives as Possible Cancer Hazards," *Science* 195, no. 4273, January 7, 1997, 17–23.

23 Andreas Sjodin, Lee-Yang Wong, Richard S. Jones, Annie Park, Yalin Zhang, Carolyn Hodge, Emily Depietro, Cheryl McClure, Wayman Turner, Larry L. Needham and Donald G. Petterson, Jr., "Serum Concentrations of Polybrominated Diphenyl Ethers (PBDEs) and Polybrominated Biphenyl (PBB) in the United States Population: 2003–2004," *Environmental Science and Technology* 42, no. 4 (February 15, 2004): 1377–84.

24 Note that the "P" in PBDE stands for "poly" and refers to a number of individual BDEs.

25 "Methyl Bromide Bill Riles 'Greens,'" *Chemical Marketing Reporter* 248, no. 1, July 3, 1995, 4.

26 Freedonia Group, *Flame Retardants to 2011*, Cleveland, August 2007.

27 Bromine Science and Environmental Forum, "An Introduction to Brominated Flame Retardants," October 19, 2000 (*http://www.ebfrip.org/download/weeeqa.pdf*). According to the BSEF, "brominated flame retardants have saved thousands of lives. In the last 10 years, a 20% reduction in fire deaths is a result of the use of flame retardants."

[28] PCB Environmental Pollution Abatement Plan. 1969. Available at www.chemicalindustryarchives.org [accessed July 5, 2008]

[29] David Rosenbaum, "Monsanto Plans to Curb Chemical," *New York Times,* July 15, 1970, 27.

[30] "Phase-Out Is Set of PCBs Chemical: Monsanto Acts after Years of Public Health Hazard," Business & Finance, *New York Times,* January 27, 1976, 54.

[31] Ami R. Zota, Ruthann A. Rudel, Rachel A. Morello-Frosch and Julian Green Brody, "Elevated House Dust and Serum Concentrations of PBDEs in California: Unintended Consequences of Furniture Flammability Standards?" *Environmental Science and Technology* 42, no. 21 (2008): 8158–64.

[32] Freedonia Group, "Executive Summary," *Flame Retardants to 2011,* Cleveland, August 2007.

Chapter 5

[1] Liz Bordo Wright, "Actress Describes Mercury Poisoning Ordeal: Daphne Zuniga Was Eating a High Seafood Diet," ABC News Program, October 21, 2005.

[2] Ibid.

[3] René Canuel, Sylvie Boucher de Grosbois, Marc Lucotte, Laura Atikessé, Catherine Larose and Isabelle Rheault, "New Evidence on the Effects of Tea on Mercury Metabolism in Humans," *Archives of Environmental and Occupational Health* 61, no. 5 (2006): 232–38.

[4] Environmental Protection Agency, "National Listing of Fish Advisories" (*http://www.epa.gov/waterscience/fish/advisories/2006/tech.html*) (accessed August 5, 2008).

[5] Oceana, "Mercury Health Effects" (*http://www.oceana.org/north-america/what-we-do/stop-seafood-contamination/the-problem/mercurys-health-effects/*) (accessed November 30, 2008).

[6] Thomas Clarkson, University of Rochester (presentation at "Mercury in the Environment" conference, University of Miami, Coral Gables, Florida, February 24, 1996).

[7] U.S. Food and Drug Administration, "An Important Message for Pregnant Women and Women of Childbearing Age Who May Become Pregnant about the Risks of Mercury in Fish," consumer advisory (http://www.cfsan.fda.gov/~dms/admehg.html) (accessed August 5, 2008).

[8] Leonardo Trasande, Philip J. Landrigan and Clyde Schechter, "Public Health and Economic Consequences of Methyl Mercury Toxicity to the Developing Brain," *Environmental Health Perspectives* 113, no. 5 (May 2005): 590–96.

[9] Clyde Johnson, "Elemental Mercury Use in Religious and Ethnic Practices in Latin American and Caribbean Communities in New York City," *Population and Environment* 20, no. 5 (May 1999).

[10] There are only a few examples of mercury being used for murder, because the slow death and telltale signs that mark its use made it less desirable than arsenic as a poison.

[11] Kaare Lund Rasmussen, Jesper Lier Boldsen, Hans Krongaard Kristensen, Lilian Skytte, Katrine Lykke Hansen, Louise Mølholm, Pieter M. Grootes, Marie-Josée Nadeau and Karen Marie Flöche Eriksen, "Mercury Levels in Danish Medieval Human Bones," *Journal of Archaeological Science* 35, no. 8 (August 2008): 2295–96.

[12] "Mercury Was Once Seen as a Cure-all," *Free Lance Star*, August 7, 2006.

[13] J.J. Putman, "Quicksilver and Slow Death," *National Geographic*, October 1972, 506–27.

[14] Tetsuya Ogura et al., "Zacatecas (Mexico) Companies Extract Hg from Surface Soil Contaminated by Ancient Mining Industries," *Water, Air, and Soil Pollution* (http://www.springerlink.com/content/100344/?p=7ba0175c831e474f98cobd 2d3b32dab4&pi=0).

[15] Pollution Probe, *Mercury in the Environment: A Primer* (Toronto: Pollution Probe, 2003).

[16] Thomas Clarkson, Laszlo Magos and Gary J. Myers, "The Toxicology of Mercury: Current Exposures and Clinical Manifestations," *New England Journal of Medicine* 349, no. 18 (October 30, 2003): 1731–37.

[17] A.C. Rennie, M. McGregor-Schuerman, I.M. Dale, C. Robinson and R. McWilliam, "Mercury Poisoning after Spillage at Home from a Sphygmomanometer on Loan from Hospital," *British Medical Journal* 319, August 7, 1999, 366–67.

[18] "Colleagues Vow to Learn from Chemist's Death," *New York Times,* October 3, 1997.

[19] Clarkson et al., "The Toxicology of Mercury."

[20] M.J. Vimy and F.L. Lorscheider, "Dental Amalgam Mercury: Background," a summary of research results on dental amalgam mercury to date, Faculty of Medicine and Medical Physiology, University of Calgary, Calgary, May 1993.

[21] Canadian Council of Ministers of the Environment, *Canada Wide Standard on Mercury for Dental Amalgam Waste,* September 2001.

[22] Philip O. Ozuah, "Mercury Poisoning," *Current Problems in Pediatrics* 30, no. 3 (March 2000): 91–99.

[23] "Are Your Teeth Toxic?" *Chicago Tribune,* December 11, 2005 (*http://www.toxicteeth.org/pressroom_articles_tribune_122005.cfm*) (accessed August 10, 2008).

[24] MedHeadlines, "Mercury Dental Fillings Not as Safe as Once Thought," June 5, 2008 (*http://medheadlines.com/2008/06/05/mercury-dental-fillings-not-as-safe-as-once-thought/*).

[25] Jillian Ashley Martin and Judy Read Guernsey, "A Comprehensive Review of the Health Effects of Fungicides," poster presentation, Department of Community Health and Epidemiology, Dalhousie University, Halifax, Nova Scotia, 2008 (*http://www.theruralcentre.com/doc/JAM-phareposter(9_29).pdf*) (accessed November 30, 2008).

[26] U.S. Centers for Disease Control, "Mercury and Vaccines (Thimerosal)" (*http://www.cdc.gov/vaccinesafety/concerns/thimerosal.htm*) (accessed November 30, 2008).

[27] Jun Ui, ed., *Industrial Pollution in Japan* (Tokyo: United Nations University Press, 1992).

[28] Tsuru Shigeto, *The Political Economy of the Environment: The Case of Japan* (Vancouver: University of British Columbia Press, 2000).

[29] Ibid. Also cited in Pollution Probe, *Mercury in the Environment.*

[30] National Institute for Minamata Disease (*http://www.nimd.go.jp/archives/english/index.html*) (accessed November 30, 2008).

[31] "Mercury Rising: The Poisoning of Grassy Narrows," *CBC News,* February 18, 2004 (*http://archives.cbc.ca/environment/pollution/topics/1178/*) (accessed August 9, 2008).

[32] G.E. McKeown-Eyssen and J. Ruedy, "Methyl Mercury Exposure in Northern Quebec: 1. Neurologic Findings in Adults," *American Journal of Epidemiology* 118, no. 4 (1983): 461–69.

[33] B. Wheatley and S. Paradis, "Exposure of Canadian Aboriginal Peoples to Methylmercury," *Water, Air, and Soil Pollution,* 1995 (*http://www.springerlink.com/content/x317026557217813/fulltext.pdf*).

[34] Pollution Probe, *Mercury in the Environment.*

[35] P. Grandjean, P. Weihe, R. White et al. "Cognitive Deficit in 7-year-old Children with Prenatal Exposure to Methylmercury," *Neurotoxicology Teratology* 19, no. 6 (1997): 417–28.

[36] Ibid.

[37] R.S. Rasmussen, J. Nettleton and M.T. Morrissey, "A Review of Mercury in Seafood: Special Focus on Tuna," *Journal of Aquatic Food Product Technology* 14, no. 1 (2005): 1–3.

Chapter 6

[1] *http://www.trumpuniversity.com/blog/post/2006/06/germiest-jobs.cfm*

[2] Environmental Working Group, "*Pesticide in Soap, Toothpaste and Breast Milk—Is it Kid-Safe?*" Washington, D.C., July 17, 2008.

[3] Antimicrobial products defend against bacteria, viruses and fungi; antibacterial products defend only against bacteria.

[4] Alliance for the Prudent Use of Antibiotics (APUA) (*www.tufts.edu/med/apua*).

[5] A.E. Aiello, B. Marshall, S.B. Levy, P. Della-Latta, S.X. Lin and E. Larson, "Antibacterial Cleaning Products and Drug Resistance," *Emerging Infectious Diseases* 11, no. 10 (October 2005): 1565–70.

[6] E.L. Larson, S.X. Lin, C. Gomez-Pichardo and P. Della-Latta, "Effect of Antibacterial Home Cleaning and Handwashing Products: Infectious Disease Symptoms," *Annals of Internal Medicine* 140, no. 5 (2004): 321–29, quoted in Aviva Glaser, "The Ubiquitous Triclosan: A Common Antibacterial Agent Exposed," *Pesticides and You* 24, no. 3 (2004).

[7] Allison E. Aiello, Elaine L. Larson and Stuart B. Levy, "Consumer Antibacterial Soaps: Effective or Just Risky?" *Clinical Infectious Diseases* 45, September 1, 2007, S146.

[8] Ibid., S144.

[9] M. Allmyr et al., "Triclosan in Plasma and Milk from Swedish Nursing Mothers and Their Exposure via Personal Care Products," *Science of the Total Environment* 372, no. 1 (2006): 87–93.

[10] Antonia M. Calafat, Xiaoyun Ye, Lee-Yang Wong, John A. Reidy and Larry L. Needham, "Urinary Concentrations of Triclosan in the U.S. Population: 2003–2004," *Environmental Health Perspectives* 116, no. 3 (2008): 303–7.

[11] D.W. Kolpin, E.T. Furlong, M.T. Meyer, E.M. Thurman et al., "Pharmaceuticals, Hormones, and Other Organic Wastewater

Contaminants in U.S. Streams, 1999–2000: A National Reconnaissance," *Environmental Science and Technology* 36 (2002): 1202–11.

[12] T.L. Miller, D.J. Lorusso, M.L. Walsh and M.L. Deinzer, "The Acute Toxicity of Penta-, Hexa- and Heptachlorohydroxydiphenyl Ethers in Mice," *Journal of Toxicology and Environmental Health* 12, nos. 2–3 (1983): 245–53. The chemical has also been found to affect the onset of puberty in a recent study: L.M. Zorrilla, E.K. Gibson, S.C. Jeffay, K.M. Crofton, W.R. Setzer, R.L. Cooper and T.E. Stoker, "The Effects of Triclosan on Puberty and Thyroid Hormones in Male Wistar Rats," *Toxicological Sciences* 107, no. 1 (January 2009): 56–64.

[13] C.M. Foran, E.R. Bennett and W.H. Benson, "Developmental Evaluation of a Potential Non-steroidal Estrogen: Triclosan," *Marine Environmental Research* 50 (2000): 153–56.

[14] Nik Veldhoen, Rachel C. Skirrow, Heather Osachoff, Heidi Waigmore, David J. Clapson, Mark P. Gunderson, Graham Van Aggelen and Caren C. Helbing, "The Bactericidal Agent Triclosan Modulates Thyroid Hormone-Associated Gene Expression and Disrupts Postembryonic Anuran Development," *Aquatic Toxicology* 80, no. 3 (December 2006): 217–27.

[15] S.B. Levy, "Antibacterial Household Products: Cause for Concern," *Emerging Infectious Diseases* 7 (2001), Suppl. no. 3: 512–15.

[16] L. Tan, N.H. Neilsen, D.C. Young and Z. Trizna, "Use of Antimicrobial Agents in Consumer Products," *Archives of Dermatology* 138, no. 8 (2002): 1082–86.

[17] Katherine Ashenburg, *The Dirt on Clean: An Unsanitized History* (Toronto: Knopf, 2007).

[18] Gunilla Sandborgh-Englund, Margaretha Adolfsson-Erici, Göran Odham and Jan Ekstrand, "Pharmacokinetics of Triclosan following Oral Ingestion in Humans: Part A," *Journal of Toxicology and Environmental Health* 69 (2006): 1861–73.

[19] Allmyr et al., "Triclosan in Plasma and Milk from Swedish Nursing Mothers," 87–93.

[20] Rebecca Sutton, *Pesticide in Soap, Toothpaste and Breast Milk—Is It Kid-Safe?* (Washington, D.C.: The Environmental Working Group, 2008).

[21] Calafat et al., "Urinary Concentrations of Triclosan," 303–7.

[22] Project on Emerging Nanotechnologies, "New Nanotech Products Hitting the Market at a Rate of 3–4 per Week," press release, April 24, 2008.

[23] Samuel N. Luomo, *Silver Nanotechnologies and the Environment: Old Problems or New Challenges,* Project for Emerging Nanotechnologies, Woodrow Wilson International Center for Scholars, Washington, D.C., September 2008.

[24] International Center for Technology Assessment, "Executive Summary," *Legal Petition Challenges EPA's Failure to Regulate Environmental and Health Threats from Nano-Silver,* International Center for Technology Assessment, Washington, D.C., May 1, 2008.

[25] "Silver Nanoparticles May Be Killing Beneficial Bacteria in Wastewater Treatment," *Science Daily,* April 30, 2008.

[26] Rye Senjen, "Nanosilver—A Threat to Soil, Water and Human Health?" Friends of the Earth Australia, Melbourne, March 2007.

[27] Ibid.

[28] Senjen, "Nanosilver—A Threat to Soil," 3.

[29] Ibid.

Chapter 7

[1] Timothy Kiely, David Donaldson and Arthur Grube, *Pesticides Industry Sales and Usage: 2000 and 2001 Market Estimates,* EPA-733-R-04-001, Environmental Protection Agency, Washington, D.C., May 2004.

[2] This is the estimated quantity of chicken wings consumed on Super Bowl weekend 2007 according to the National Chicken Council (Reuters, January 22, 2008 [http://www.reuters.com/article/pressRelease/idUS186633+22-Jan-2008+PRN20080122]).

[3] Caroline Cox, "2,4-D: Toxicology," *Journal of Pesticide Reform* 19, no. 1 (Spring 1999).

[4] Wolfgang Mieder, "The Grass Is Always Greener on the Other Side of the Fence: An American Proverb of Discontent," 1995 (*www.deproverbio.com*).

[5] Environmental Protection Agency, *DDT Regulatory History: A Brief Survey (to 1975)*, July 1975.

[6] Linda Lear, *Rachel Carson: Witness for Nature* (New York: Henry Hoyten, 1997).

[7] Edmund Russell, *War and Nature: Fighting Humans and Insects with Chemicals from World War I to* Silent Spring (Cambridge: Cambridge University Press, 2001).

[8] Ibid.

[9] "Industry Task Force II on 2,4-D Research Data," December 2005 (*www.24d.org*).

[10] James Troyer, "In the Beginning: The Multiple Discovery of the First Hormone Herbicides," *Weed Science* 49, Issue 2 (March-April 2001): 290–97.

[11] "Taking Stock of DDT," *American Journal of Public Health* 36, no. 6 (June 1946).

[12] Russell, *War and Nature*.

[13] Kelli Miller Stacy, "Cancer Risk Lingers for Long-Banned DDT," *WebMD*, April 29, 2008 (*http://men.webmd.com/news/20080428/cancer-risk-lingers-for-long-banned-ddt*).

[14] Michel Aubé, Christian Larochelle and Pierre Ayotte, "1,1-dichloro-2,2-bis(p-chlorophenyl)ethylene(p,p'-DDE) Disrupts the Estrogen-Androgen Balance Regulating the Growth of Hormone-Dependent Breast Cancer Cells," *Breast Cancer Research* 10, no. 1 (February 2008), quoted in "DDT Compound Speeds Breast Cancer Growth," HealthDay News, *MedicineNet*, February 14, 2008 (*http://www.medicinenet.com/script/main/art.asp?articlekey=87155*).

[15] Caroline Cox, "2,4-D Toxicology: Part 2," *Journal of Pesticide Reform* 19, no. 2 (Summer 1999).

[16] The testing was conducted by Accu-Chem Laboratories, Richardson, Texas.

[17] Chensheng Lu et al., "Organic Diets Significantly Lower Children's Dietary Exposure to Organophosphorus Pesticides," *Environmental Health Perspectives* 114, no. 2 (February 2006).

[18] LL. Needham, V.W. Burse, S.L. Head et al., "Adipose Tissue/Serum Partitioning of Chlorinated Hydrocarbon Pesticides in Humans," *Chemosphere* 20 (1990): 975–80.

[19] United Nations Environment Programme, *Central America and the Caribbean: Regional Report,* December 2002.

[20] Centers for Disease Control and Prevention, *Third Report on Human Exposure to Environmental Chemicals,* Atlanta, 2005.

[21] Environmental Working Group, Human Toxome Project (*http://www.bodyburden.org/*).

[22] Ibid.

[23] Marsha K. Morgan, Linda S. Sheldon, Kent W. Thomas, Peter P. Egeghy, Carry W. Croghan, Paul A. Jones, Jane C. Chuang and Nancy K. Wilson, "Adult and Children's Exposure to 2,4-D from Multiple Sources and Pathways," *Journal of Exposure Science and Environmental Epidemiology* 18, no. 5 (2008): 486–94.

[24] Ibid.

[25] Kiely et al., *Pesticides Industry Sales and Usage.*

[26] Environmental Protection Agency, *2,4-D RED Facts,* EPA-738-F-05-00, Washington, D.C., June 30, 2005 (*http://www.epa.gov/oppsrrd1/REDs/factsheets/24d_fs.htm*).

[27] G.R. Stephenson, K.R. Solomon and L. Ritter, "Environmental Persistence and Human Exposure Studies with 2,4-D and Other Turfgrass Pesticides," Centre for Toxicology, University of Guelph (*http://www.envbio.uoguelph.ca/pdf/persistence.pdf*).

[28] David Boyd, *Unnatural Law: Rethinking Canadian Environmental Law and Policy* (Vancouver: University of British Columbia Press, 2003).

[29] Dr. Sheila Basrur was the Medical Officer of Health for the City of Toronto at the time. She went on to become Ontario's Chief Medical Officer but died of cancer in 2008. She was one of Canada's greatest leaders in protecting public health.

[30] M. Christie, *Private Property Pesticide By-laws in Canada*, Coalition for a Healthy Ottawa, May 1, 2008 (*http://www.flora.org/healthyottawa/BylawList.pdf*).

[31] Statistics Canada, "Households and the Environment Survey," *Daily*, July 11, 2007 (*http://www.statscan.ca/Daily/English/070711/d070711b.htm*).

[32] "Dow Contests Pesticide Ban," *Chemical and Engineering News* 86, no. 44, November 22, 2008.

[33] K.L. Bassil, C. Vakil, M. Sanborn, D.C. Cole, J.S. Kaur and K.J. Kerr, "Cancer Health Effects of Pesticides: Systematic Review," *Canadian Family Physician* 53, no. 10 (2007): 1704–11.

[34] Tye E. Arbuckle et al. "2,4-Dichlorophenoxyacetic Acid Residues in Semen of Ontario Farmers," *Reproductive Toxicology* 13 (1999): 421–29.

[35] G.M. Solomon and P.M. Weiss, "Chemical Contaminants in Breast Milk: Time Trends and Regional Variability," *Environmental Health Perspectives* 110, no. 6 (2002): A336–A347.

[36] K.L. Bassil, C. Vakil, M. Sanborn, D.C. Cole, J.S. Kaur and K.J. Kerr, "Cancer Health Effects of Pesticides: Systematic Review," *Canadian Family Physician* 53, no. 10 (2007): 1704–11.

[37] M. Sanborn, K.J. Kerr, L.H. Sanin, D.C. Cole, K.L. Bassil and C. Vakil, "Non-cancer Health Effects of Pesticides: Systematic Review and Implications for Family Doctors," *Canadian Family Physician* 53 (October 2007).

[38] Ibid.

[39] Dow AgroSciences, "About Us" (*http://www.dowagro.com/about/*) (accessed December 17, 2008).

[40] M. Sears, C.R. Walker, R.H.C. van der Jagt and P. Claman, "Pesticide Assessment: Protecting Public Health on the Home Turf," *Pediatrics and Child Health* 11, no. 4 (April 2006): 229–34.

[41] Theo Colborn, "A Case for Revisiting the Safety of Pesticides: A Closer Look at Neurodevelopment," *Environmental Health Perspectives* 114 (2006): 10–17.

[42] Sheldon Rampton and John Stauber, *Trust Us, We're Experts: How Industry Manipulates Science and Gambles with Your Future* (New York: Tarcher/Putnam, 2002).

[43] *http://www.teflon.com/Teflon/teflonissafe/teflonissafe.html*

[44] Suemedha Sood, "Scientists: EPA 'Under Siege' Survey of EPA Scientists Finds Rampant Political Interference," *Washington Independent*, April 23, 2008 (*http://washingtonindependent.com/1577/scientists-epa-under-siege*).

[45] "Global Top 50," *Chemical and Engineering News* 85, no. 30 (July 23, 2007) (*http://pubs.acs.org/cen/coverstory/85/8530cover.html*).

[46] Auditor General of Canada, "Managing the Safety and Accessibility of Pesticides," chap. 1 in *Report of the Commissioner of the Environment and Sustainable Development* (Ottawa: Office of the Auditor General of Canada, 2003).

[47] "Review Points to Holes in CFIA Food System," *CBC News*, September 24, 2008.

[48] William Leiss, *In the Chamber of Risks: Understanding Risk Controversies* (Montreal and Kingston: McGill–Queen's University Press, 2001).

[49] Oracle Poll Research, *Survey Report Prepared for Pesticide Free Ontario and the Canadian Association of Physicians for the Environment*, February 2007.

Chapter 8

[1] Rick Smith and Adam Daifallah, "It's in Tory Genes to Go Green," *Globe and Mail*, February 6, 2006.

[2] Terence Corcoran, "Rona Brockovich?" *National Post*, June 3, 2006.

[3] Antonia M. Calafat, Xiaoyun Ye, Lee-Yang, John A. Reidy and Larry Needham, "Exposure of the U.S. Population to Bisphenol A and 4-tertiary-

octylphenol: 2003–2004," *Environmental Health Perspectives* 116, no. 1 (January 2008): 39–44.

4 Steve Toloken, "SPI Study Disputes Endocrine Disruptor Findings," *Plastics News,* October 16, 1998.

5 Notes within Table 7 are as follows: (i) C.M. Markey, P.R. Wadia, B.S. Rubin, C. Sonnenschein and A.M. Soto, "Long-Term Effects of Fetal Exposure to Low Doses of the Xenoestrogen Bisphenol-A in the Female Mouse Genital Tract," *Biology of Reproduction* 72, no. 6 (2005): 1344–51; (ii) M. Muñoz-de-Torono, C.M. Markey and P.R. Wadia, "Perinatal Exposure to Bisphenol-A Alters Peripubertal Mammary Gland Development in Mice," *Endocrinology* 146, no. 9 (2005): 4138–87; (iii) V. Bindhumol, K.C. Chitra and P.P. Mathur, "Bisphenol A Induced Reactive Oxygen Species Generation in the Liver of Male Rats," *Toxicology* 188, nos. 2–3 (2003): 117–24; (iv) S.C. Nagel, F.S. vom Saal, K.A. Thayer, M.G. Dhar, M. Boechler and W.V. Walshons, "Relative Binding Affinity-Serum Modified Access (RBA-SMA) Assay Predicts the Relative in Vivo Bioactivity of the Xenoestrogens Bisphenol-A and Octylphenol," *Environmental Health Perspectives* 105, no. 1 (1997): 70–76; (v) S. Homna, A. Suzuki, D.L. Buchanan, Y. Katsu, H. Watanabe and T. Iguchi, "Low-Dose Effect of in Utero Exposure to Bisphenol A and Diethylstilbestrol on Female Mouse Reproduction," *Reproductive Toxicology.* 16, no. 2 (2002): 117–22; (vi) B.T. Akingbemi, C.M. Sottas, A.I. Loulova, G.R. Klinefelter and M.P. Hardy, "Inhibition of Testicular Steroidogenesis by the Xenoestrogen Bisphenol A Is Associated with Reduced Pituitary Luteinizing Hormone Secretion and Decreased Steroidogenic Enzyme Gene Expression in Rat Leydig Cells," *Endocrinology* 145, no. 2 (2004): 592–603; (vii) T.J. Murray, M.V. Maffini, A.A. Ucci, C. Sonnenschein and A.M. Soto, "Induction of Mammary Gland Ductal Hyperplasias and Carcinoma in Situ following Fetal Bisphenol A Exposure," *Reproductive Toxicology* 23, no. 3 (2007): 383–90; (viii) S.M. Ho, W.Y. Tang, J. Belmonte de Frausto and G.S. Prins, "Developmental Exposure to Estradiol and Bisphenol A Increases Susceptibility to Prostate Carcinogenesis and Epigenetically Regulates

Phosphodiesterase Type 4 Variant 4," *Cancer Research,* 66, no. 11 (2006): 5624–32; (ix) P.L. Palanza, K.L. Howdeshell, S. Parmigiani and F.S. vom Saal, "Exposure to a Low Dose of Bisphenol A during Fetal Life or in Adulthood Alters Maternal Behavior in Mice," *Environmental Health Perspectives,* 110 (2002), Suppl. no. 3: 415–22; (x) K. Kubo, O. Arai, M. Omura, R. Watanabe, R. Ogata and S. Aou, "Low Dose Effects of Bisphenol A on Sexual Differentiation of the Brain and Behavior in Rats," *Neuroscience Research,* 45, no. 3 (2003): 345–56; (xi) Csaba Leranth, Tibor Hajszan, Klara Szigeti-Buck, Jeremy Bober and Neil J. MacLusky, "Bisphenol A Prevents the Synaptogenic Response to Estradiol in Hippocampus and Prefrontal Cortex of Ovariectomized Nonhuman Primates," *Proceedings of the National Academy of Sciences* 105, no. 37 (2008): 14187–91; (xii) Eric R. Hugo, Terry D. Brandebourg, Jessica G. Woo, Jean Loftus, J. Wesley Alexander and Nira Ben-Jonathan, "Bisphenol A at Environmentally Relevant Doses Inhibits Adipoentectin Release from Human Adipose Tissue Explants and Adipocytes," *Environmental Health Perspectives* (2008) (accessed online August 14, 2008); (xiii) Environmental Protection Agency, *Oral RfD Assessment: Bisphenol A Integrated Risk Information System,* Washington, D.C., 1998 (*http://www.epa.gov/iris/subst/0356.htm.*); (xiv) P. Alonso-Magdalena, A.B. Ropero, M.P. Carrera, C.R. Cederroth, M. Bacquié, B.R. Gautheir, S. Nef, E. Stefani and A. Nadal, "Pancreatic Insulin Content Regulation by the Estrogen Receptor ER Alpha," *PLoS ONE* 3, no. 4 (2008): e2069.

[6] F.S. vom Saal, B.T. Akingbemi, S.M. Belcher, L.S. Birnbaum, D.A. Crain, M. Eriksen, F. Farabollini, L.J. Guillette, R. Hauser, J.J. Heindel, S.M. Ho, P.A. Hunt, T. Iguchi, S. Jobling, J. Kanno, R.A. Keri, K.E. Knudsen, H. Laufer, G.A. LeBlanc, M. Marcus, J.A. McLachlan, J.P. Myers, A. Nadal, R.R. Newbold, N. Olea, G.S. Prins, C.A. Richter, B.S. Rubin, C. Sonnenschein, A.M. Soto, C.D. Talsness, J.G. Vandenbergh, L.N. Vandenberg, D.R. Walser-Kuntz, C.S. Watson, W.V. Welshons, Y. Wetherill and R.T. Zoeller, "Chapel Hill Bisphenol A Expert Panel Consensus Statement: Integration of Mechanisms, Effects in Animals and Potential to Impact Human Health at Current Levels of Exposure," *Reproductive Toxicology* 24, no. 2 (2007): 131–38.

[7] Monica Muñoz-de-Toro, Caroline M. Markey, Perinaaz R. Wadia, Enrique H. Luque, Beverly S. Rubin, Carlos Sonnenschein and Ana M. Soto, "Perinatal Exposure to Bisphenol-A Alters Peripubertal Mammary Gland Development in Mice," *Endocrinology* 146, no. 9, September 1, 2005, 4138–47 (*http://endo.endojournals.org/cgi/content/full/146/9/4138*).

[8] A container which, in a recent study, was found to leach bisphenol A: *http://dev.www.jsonline.com/watchdog/watchdogreports/34532859.html*

[9] *http://safemama.com/about/* (accessed December 28, 2008).

Chapter 9

[1] Bliss Shepherd, "Scott Nearing on War," *Counterpunch*, Petrolia, California November 6, 2001.

[2] Environmental Protection Agency, Office of Air and Radiation, *Report to Congress on Indoor Air Quality*, vol. 2, *Assessment and Control of Indoor Air Pollution*, EPA 400-1-89-001C, Washington, D.C., 1989, I, 4–14.

[3] Both of us try to avoid toxins in our daily lives, and we make conscious choices to limit our exposure. Yet even before embarking on our toxic experimentation, Rick had measurable levels of five phthalates, nine PCBs, eight PBDEs, triclosan, bisphenol A and six PFCs—and Bruce was polluted with mercury, six PFCs and five pesticides. And these were just the chemicals we tested for. Without a doubt if we had run the full gamut of expensive tests currently available, we would have turned up hundreds of toxins in our bodies.

[4] Alison Cohen, Sarah Janssen and Gina Solomon, *Clearing the Air: Hidden Hazards of Air Fresheners*, Natural Resources Defense Council issue paper, September 2007.

[5] Jeff Johnson, "A Tsunami of Electronic Waste: No High-Tech Solution for the Detritus of the Information Revolution," *Chemical and Engineering News* 86, no. 21 (May 26, 2008): 32–33.

[6] Environment Canada, *EnviroZine* 33, June 26, 2003 (*http://www.ec.gc.ca/envirozine/english/issues/33/feature1_e.cfm*).

CREDITS

Grateful acknowledgment is made to the following for the permission to reprint previously published material:

Quotation on page 8 is from *Silent Spring* by Rachel Carson. © 1962 by Rachel L. Carson, renewed 1990 by Roger Christie. Reprinted by permission of Houghton Mifflin Harcourt Publishing Company. All rights reserved; Quotation on page 9 is from, *Society and Solitude: Twelve Chapters* by Ralph Waldo Emerson. The Edwin Mellen Press, Lewiston, NY, 2008. Reprinted by permission; Quotation on page 15 is from "Arabic Treatises on Environmental Pollution up to the End of the Thirteenth Century" by Lutfallah Gari, published in *Environment and History*, published by The White Horse Press. © 2002 Reprinted by permission; Quotation on page 16 is from *The Condition of the Working Class in England in 1844* by Friedrich Engels (1993). By permission of Oxford University Press; Quotation on page 20 is from Consumers Union of U.S., Inc. Yonkers, NY 10703-1057, a nonprofit organization. Reprinted with permission from ConsumerReports.org®. Illustration No. 1 on page 23 is from "*Is Mother's Milk Fit for Human Consumption?*" by Environmental Defense Fund, published in the *New York Times* 1970. Reprinted by permission; Quotation on page 24 is from "50,000 Baby Teeth" by W.K. Wyant., Jr., published in *The Nation*, June 1959. Reprinted by permission; Illustration No. 2 on page 25 is from the Western Historical Manuscript Collection. Reprinted by permission; Quotation on page 33 is from Rita Rudner, Comedian,

Ritmar Productions. Reprinted by permission; Quotation on page 39 is from "Phthalate Monoester Levels in the Urine of Young Children" by J.W. Brock, S.P. Caudill, M.J. Silva, L.L. Needham, and E.D. Hilborn, published in *Bulletin of Environmental Contamination and Toxicology* Vol 68 (2002): 309–14. © 2002 Springer-Verlag New York Inc. Reprinted with kind permission of Springer Science and Business Media; Quotation on page 42 is from "Pediatric Exposure and Potential Toxicity of Phthalate Plasticizers" by Katherine M. Shea and Committee on Environmental Health, published in *Pediatrics Journal*. Reprinted by permission from the American Academy of Pediatrics; Quotation on page 44 is from "Consumer Probe," *Saturday Night Live*, Episode 10, Season 2, November 12, 1976, published by NBC Courtesy of Broadway Video Enterprises and NBC Studios, Inc. © 1976 NBC Studios Inc. Distributed by Broadway Video Enterprises; Illustration No. 8 on page 58 "No Yucky in My Duckie" permission granted by Genevieve K. Howe, Ecology Center and Sara Talpos (parent of Jackson, the child in the photo); Illustration No. 9 on page 58 "Save the Rubber Duckies" by Consumers for Competitive Choice; Quotation on page 69 is from "Shimmer," *Saturday Night Live*, Episode 9, Season 1, January 10, 1976, published by NBC Courtesy of Broadway Video Enterprises and NBC Studios, Inc. © 1976 NBC Studios Inc. Distributed by Broadway Video Enterprises; Quotation on page 72 is from CBC *Marketplace*: "PFOA: What is it, and how did it get into our blood?" Broadcast March 20 2005. Excerpted from the interview with Erica Johnson, *Marketplace* reporter, and Scott Mabury, an environmental chemist at the University of Toronto; Quotation on page 75 is from interview with Della Tennant, National Public Radio, January 2006; Quotation on page 84 is from "First C8 Study Results Listed" by Pamela Brust, published in the *Parkersburg News and Sentinel*, October 2008; Quotation on page 96 is from "Cat People (Putting Out Fire)" words by David Bowie, music by Giorgio Moroder. © 1982 Universal Music Corp. and Songs of Universal, Inc. Used by permission; Quotation on page 100 is from "Great Lakes Chemical: An Aggressive Growth Plan Pays Off," published in *Chemical Week*, March 21, 1984. Reprinted by permission; Quotation on page 102 is from "Intentions Gone Astray: The Facts about Tris Don't Leave Much Choice" by Linda Charlton, published in the *New York Times*, The Week in Review, July 3, 1977. Reprinted by permission; Quotation on page 102 is from the Environmental Defense Fund Petition; Quotations on page 102 are from "Flame Retardant Sleepwear: Is There a Risk of Cancer?" by Nadine Brozan, published in the *New York Times*, Travel Section-April 10, 1976. Reprinted by permission; Quotation on page 107 is from "Safety of a New Flame Retardant Questioned" by Joan Lowy, published in Scripps Howard News Service, 2004. Reprinted by permission; Quotations on page 131 are from "Actress Describes Mercury Poisoning Ordeal" written by Liz Bordo Wright, published by ABC News Program; Quotation on page 159 is from Donald Trump, published on the Donald Trump University website, 2006; Quotation on page 179 is from the *Oxford Dictionary*, published by Market House Books Inc. Reprinted by permission; Illustration No. 19 on page 181 is from the Project on Emerging Nanotechnologies 2006. Reprinted by permission; Quotation on page 183 is from "Nanosilver: A Threat to Soil, Water and Human Health?" by Rye Senjen, published in *Friends of the Earth Australia*, 2007. Reprinted by permission; Quotation on page 189 is from "In the Beginning: The Multiple Discovery of the First Hormone Herbicides" by James Troyer, published in *Weed Science*, 2001. © 2008 Weed Science Society of America/Allen Press Publishing Services; Quotation on page 197 is from *Environmental Persistence and Human Exposure Studies with 2,4-D and Other Turfgrass Pesticides* by G.R. Stephenson, K.R. Solomon and L. Ritter, published by Centre for Toxicology, University of Guelph; Quotation on page 210 is from "Scientists: EPA 'Under Siege' Survey of EPA Scientists Finds Rampant Political Interference" by Suemedha Sood, published by the *Washington Independent*, 2008. Reprinted by permission; Quotation on page 214 is from *In the Chamber of Risks: Understanding Risk Controversies* by William Leiss, published by McGill-Queen's University Press,

2001; Quotation on page 216 is from *The Graduate*, 1967, published by Vivendi/Canal © 1967. Studio Canal; Quotation on page 226 is from "Rona Brockovich?" by Terence Corcoran, published in the *National Post*, 2006. Reprinted by permission; Illustration No. 22 on page 218 BPA rally photo of young girl holding her baby picket. Permission to reprint by Monique Fabregas; Quotation on page 237 is from "Chapel Hill Bisphenol A Expert Panel Consensus Statement: Integration of Mechanisms, Effects in Animals and Potential to Impact Human Health at Current Levels of Exposure." By F.S. vom Saal, B.T. Akingbemi, S.M. Belcher, L.S. Birnbaum, D.A. Crain, M. Eriksen, F. Farabollini, L.J. Guillette, R. Hauser, J.J. Heindel, S.M. Ho, P.A. Hunt, T. Iguchi, S. Jobling, J. Kanno, R.A. Keri, K.E. Knudsen, H. Laufer, G.A. LeBlanc, M. Marcus, J.A. McLachlan, J.P. Myers, A. Nadal, R.R. Newbold, N. Olea, G.S. Prins, C.A. Richter, B.S. Rubin, C. Sonnenschein, A.M. Soto, C.E. Talsness, J.G. Vandenbergh, L.N. Vandenberg, D.R. Walser-Kuntz, C.S. Watson, W.V. Welshons, Y. Wetherill, R.T. Zoeller, published by *Reproductive Toxicology* 2007. Copyright © 2007, published by Elsevier Inc. Reprinted by permission; Illustration No. 24 on page 239 is from "Perinatal Exposure to Bisphenol-A Alters Peripubertal Mammary Gland Development in Mice" by Monica Munoz-de-Toro, Caroline M. Markey, Perinaaz R. Wadia, Enrique H. Luque, Beverly S. Rubin, Carlos Sonnenschein and Ana M. Soto. *Endocrinology* Vol. 146, No. 9 4138-4147 Copyright 2005, The Endocrine Society. Reprinted by permission; Quotation on page 254 is from the song "Substitute" by Pete Townshend, The Who, from the album *Meaty, Beaty and Bouncy*, published by DEVON MUSIC INC. Reprinted by permission.

Every effort has been made to contact the copyright holders; in the event of an inadvertent omission or error, please notify the publisher.

INDEX

Page numbers in italics indicate illustrations.

As Executive Director of Environmental Defence Canada, RICK SMITH is one of that country's leading environmentalists. He holds a doctorate in biology from the University of Guelph.

BRUCE LOURIE, an influential environmental thinker, started one of Canada's largest environmental consultancies. He works closely with governments, businesses, foundations and non-profit organizations. He is President of the Ivey Foundation.

SARAH DOPP is a veteran grassroots organizer, political staffer and campaigner.

ACKNOWLEDGMENTS

FIRST AND FOREMOST, this unusual project would not have been possible were it not for the support and love of our families. Rick would like to thank his wife, Jennifer Story, for her patience, wise counsel and duck-themed book title idea; his two little boys, Zachary and Owain, for making him bust a gut laughing and reminding him what's truly important in life; and Dave Lavigne, for his mentorship and for passing on the liberating insight that you can begin sentences with "And." Bruce thanks Biz, Ellen and Claire, who missed more than a few home-cooked meals throughout the writing process, and the Ivey Foundation. Sarah would like to thank David Noble.

The inspiration for this book arose from Environmental Defence Canada's groundbreaking *Toxic Nation* campaign. This multiyear effort was made possible through the dedication of the staff and Board of Environmental Defence, especially Jennifer Foulds, Aaron Freeman, Kapil Khatter, Jana Neumann, Cassandra Polyzou, and Sarah Winterton.

We would like to thank the many people who gave generously of their time to review our experimental protocols and manuscript: Katherine Ashenburg, Åke Bergman, Craig Butt, Ken Cook, Susan Duty, Coreen Hamilton, Jamey Heath, Jane Houlihan, Pat Hunt, Stuart Levy, Marc Lucotte, Scott Mabury, Mike Matisko, Burkhard Mausberg, Pete Myers, Janet Nudelman, Gail Prins, Deborah Rice, Ted Schettler, Heather Stapleton, Shanna Swan, Cathy Vakil, Fred vom Saal, Julia Taylor and Tom Webster. The leadership of Pete Myers and of the Environmental Working Group, in particular, has shaped our thinking regarding toxic chemicals and their place in our lives. Any deficiencies in the text of the book are entirely our own, not theirs.

Invaluable research assistance was provided by Louise Pilfold, and Lorraine Johnson provided important advice. Our gratitude to Knopf Canada's Louise Dennys for believing in this project, to our editor Michael Schellenberg for his understanding as we juggled the book and our day jobs, to Kathryn Dean for some dynamite copy editing and to Michelle MacAleese for helping keep things on the rails. Thanks to the team of Jack Shoemaker, Charlie Winton, Sharon Donovan, Tiffany Lee and Laura Mazer at Counterpoint Press.

Karee Dryden, nurse extraordinaire, drew a mean blood sample. Accu-Chem Laboratories (Richardson, Texas), Axys Analytical Services (Sidney, British Columbia), Brooks Rand Labs (Seattle, Washington) and STAT Analysis Corporation (Chicago, Illinois) analyzed our blood, urine and toy samples in record time, and we thank them for their efficiency and professionalism.

Finally, we want to acknowledge the many enlightened public officials, journalists, scientific researchers, community leaders, parents and other engaged citizens who—often in the face of very long odds—take action every day to protect our environment and health. Of particular note is Dr. Sheila Basrur, who died of cancer in 2008 and was never afraid to step out and lead efforts to protect public health.

Together, we are winning.